AF497264

Introduction to the Geometrical Foundations of General Relativity

A Compendium of Tensor Calculus for Physicists

Bernd Wichmann

2nd edition

Bernd Wichmann, Kufstein, Austria, bernd47@kufnet.at, www.tensor-calculus.com

To Helga

Preface

This book is intended for physics students (undergraduate/transition to graduate) who want to prepare for lectures on general relativity. Some knowledge of linear algebra and analysis are required. This textbook starts with basic topics such as vector space and vectors (chapter 1), dual space and covectors (chapter 2), tensors (chapter 3), etc.. Great importance is always attached to the clarity of the explanations and derivations of the topics. 31 Figures support these intentions.

General relativity is in its deeper sense a geometric theory. Therefore, the emphasis in this textbook has been placed on understanding space in its geometric configuration. Space is a component of the representation of the physical real. And it is thus the stage on which the physical processes and procedures show themselves.

Calculation tasks have been deliberately omitted. The focus is on understanding a topic. For this purpose, many examples and detailed extra introductions have been made. For exercises, there are enough examples in the relevant textbooks that can be used to deepen a topic. This book serves as a good basis for mastering tasks. If you have studied the book thoroughly, you will be prepared to start working on the physics of gravitation as described by general relativity.

Acknowledgments

Text and mathematical formulas were written with LaTeX. The Figures were created with the programme mathematica® from Wolfram Research, Inc. I would like to thank Simone Szurmant of ADDITIVE, Friedrichsdorf, Germany, and Thomas Ponweiser of uni software plus, Linz, Austria, for their support with problems and questions concerning mathematica®.

Contents

CONTENTS

List of Figures

Chapter 1

Vector Space and Vectors

1.1 Vector Space

A *vector space* $\mathcal{V}$ contains the following arithmetic operations over its elements: addition and scaling. For our geometric investigations we call its elements *vector*. Vectors are drwan as arrows with value (length) and direction to make it geometrically visual, Figure 1.1 and Figure 1.2. There are also vector spaces over polynomials, matrices, tensors etc.

Let $\vec{u}, \vec{v}$ and $\vec{w}$ be vectors in $\mathcal{V}$, a and b real numbers, then the following *axioms* apply to a real vector space $\mathcal{V}$:

1. $\vec{v} + \vec{w} = \vec{w} + \vec{v} \in \mathcal{V}$,

2. $\vec{u} + (\vec{v} + \vec{w}) = (\vec{u} + \vec{v}) + \vec{w} \in \mathcal{V}$,

3. $a(\vec{v} + \vec{w}) = a\vec{v} + a\vec{w} \in \mathcal{V}$,

4. $(a + b)\vec{v} = a\vec{v} + b\vec{v} \in \mathcal{V}$,

5. $(ab)\vec{v} = a(b\vec{v}) \in \mathcal{V}$,

6. $1 \cdot \vec{v} = \vec{v} \in \mathcal{V}$.

The identity element of $\mathcal{V}$ is the *zero vector* $\vec{0}$: $\vec{v} + \vec{0} = \vec{0} + \vec{v} = \vec{v}$.

1.2 Basis

In order to be able to operate calculus in a vector space, one introduces basis vectors which span the entire vector space $\mathcal{V}$. The number of necessary basis vectors corresponds to the dimension n of the vector space. The algebraic relationship for an ordered set of basis vectors $\{\vec{e}_1, \vec{e}_2, \ldots, \vec{e}_n\}$ and a set $\{a_i\}$ of real numbers is:

$$a^1\vec{e}_1 + a^2\vec{e}_2 + \cdots + a^n\vec{e}_n = \vec{0}, \text{ only for } a^i = 0, \forall i. \tag{1.1}$$

This equation has *no* non-trivial solution. Basis vectors are thus *linearly independent* . No basis vector is a linear combination of other basis vectors.

Remark 1.1

You can of course define different bases for an n-dimensional vector space. Example: Cartesian, polar, etc.

■

Each vector $\vec{v}$ is *uniquely determined* by the coefficients a^i using the basis vectors. These coefficients are called the components of the vector.

$$\vec{v} = a^1\vec{e}_1 + a^2\vec{e}_2 + \cdots + a^n\vec{e}_n.$$

Proof:

We assume that there is a second representation of vector $\vec{v}$:

$$\vec{v} = b^1\vec{e}_1 + b^2\vec{e}_2 + \cdots + b^n\vec{e}_n.$$

We form the difference of the vector to itself, obtaining the zero vector

$$\begin{aligned}
\vec{0} &= \vec{v} - \vec{v} \\
&= (a^1\vec{e}_1 + a^2\vec{e}_2 + \cdots + a^n\vec{e}_n) - (b^1\vec{e}_1 + b^2\vec{e}_2 + \cdots + b^n\vec{e}_n) \\
&= (a^1 - b^1)\vec{e}_1 + (a^2 - b^2)\vec{e}_2 + \cdots + (a^n - b^n)\vec{e}_n,
\end{aligned}$$

and because of the linear independence of the basis vectors $\vec{e}_i$, equation (1.1) applies: $(a^i - b^i) = 0.$

■

1.3 Components

A vector $\vec{v}$ in a 2-dimensional vector space $\mathcal{V}$ with basis vectors $\vec{e}_i$ can be described algebraically as follows:

$$\vec{v} = v^1 \vec{e}_1 + v^2 \vec{e}_2. \tag{1.2}$$

The numbers v^i are called the *components* of the vector $\vec{v}$. In linear algebra, they are summarized in a *column vector*:

$$\vec{v} = \begin{pmatrix} v^1 \\ v^2 \end{pmatrix} \tag{1.3}$$

Remark 1.2
As stated in Section 1.1 in this textbook we work only with vector spaces based on the real numbers $\mathbb{R}$ (components, scalars, functions, etc.).

$\blacksquare$

A position vector $\vec{R}$ is a directional vector from the origin of the coordinate system to a selected point of the vector space. The Figure 1.1 shows an example of a position vector $\vec{R}$ in a 2-dimensional coordinate system. Once in a Cartesian coordinate system and the other in a general coordinate system. In contrast to a position vector $\vec{R} = 2 \cdot \vec{e}_1 + 1 \cdot \vec{e}_2$ (Figure 1.1), the vector

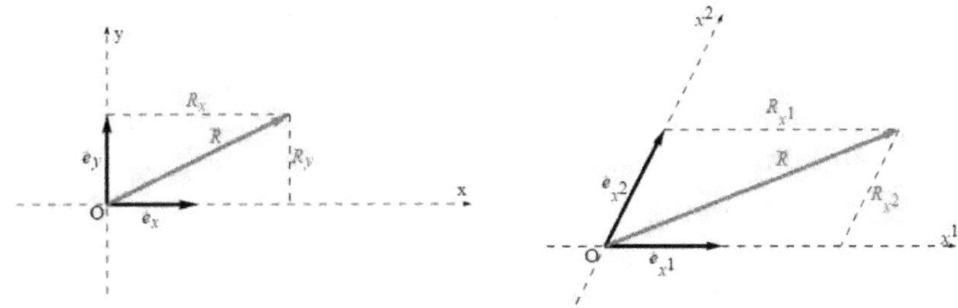

Figure 1.1: Position vector $\vec{R}$ in two different coordinate systems.

$\vec{v} = 2 \cdot \vec{e}_1 + 1 \cdot \vec{e}_2$ represents a *vector field*. See Figure 1.2.

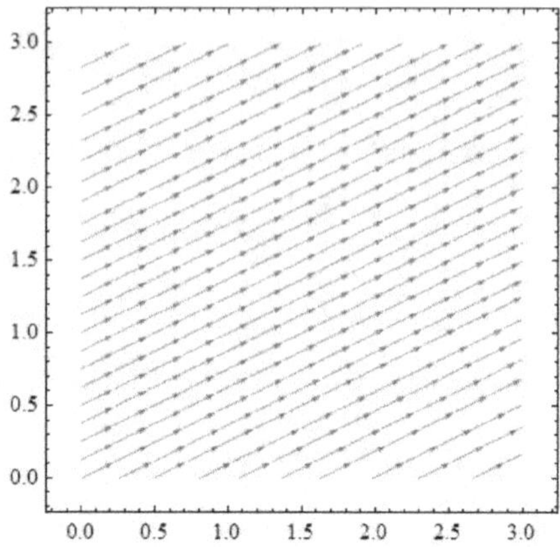

Figure 1.2: Vector field $\vec{v} = 2 \cdot \vec{e}_1 + 1 \cdot \vec{e}_2$.

Remark 1.3

The *components* of a vector $\vec{v}$ of a vector space $\mathcal{V}$ are always provided with a *superscript* (upstairs, upper) index, the *basis vectors* $\vec{e}_i$ of a vector space $\mathcal{V}$ always with a subscript (downstairs, lower) index. In the dual space $\mathcal{V}^\star$ (see Chapter 2), the situation is reversed.

■

Box 1.1

We mostly use the *Einstein summation convention*:

$$\sum_{i=1}^{n} v^i \vec{e}_i = v^i \vec{e}_i, \quad \sum_{k} a^k b_k, \text{ and so on.} \tag{1.4}$$

In other words, identical superscripts and subscripts are added up. This is the so called *contraction* . See later sections.

Of course, the summation indices can be supplemented by other symbols (*dummy indices*) . But attention: already used letters, symbols in the term to be edited may **not** be reused!

$$a^k b_k = a^i b_i = a^\spadesuit b_\spadesuit = \cdots,$$
$$\text{but:} \quad a^k b_k c^i \neq a^i b_i c^i$$

Not summed up indices are called *free* indices . ■

1.4 Transformation

In this text, mainly the *coordinate basis* is used, which is expressed by means of the partial differential operator, applied to the coordinate system $\{x^i\}$:

$$\vec{e}_k \equiv \frac{\partial}{\partial x^k} \equiv \partial_k. \tag{1.5}$$

Partial derivatives and coordinate basis vectors as defined by equation (1.5) are the same because $\frac{\partial}{\partial x^k}$ follows the x^k coordinate line. An important property of a coordinate basis is that its commutator vanishes (partial derivatives commutate).

$$\left[\frac{\partial}{\partial x^i}, \frac{\partial}{\partial x^j} \right] = [\partial_i, \partial_j] = 0. \text{ No restrictions for } i \text{ and } j. \tag{1.6}$$

Remark 1.4
It should be noted that in general the basis vectors $\vec{e}_k$ of a coordinate basis, equation (1.5), are *not* normalized.

■

Box 1.2
An essential feature of a coordinate system is that coordinate x^i is constant along coordinate x^j, $i \neq j$. Thus the relation

$$\frac{\partial x^i}{\partial x^j} = \delta^i_j = \begin{cases} 1 & i = j \\ 0 & i \neq j \end{cases}$$

applies, which we will repeatedly encounter in this textbook.
The *Kronecker delta* is written $\delta_{ij}, \delta^i_j, \delta^{ij}$ and is defined by

$$\delta_{ij}, \delta^i_j, \delta^{ij} = \begin{cases} 1 & i = j, \\ 0 & i \neq j. \end{cases} \tag{1.7}$$

As an example we treat the transition from Cartesian coordinates to polar coordinates and vice versa. It is known:

$$x = r\cos\theta, y = r\sin\theta. \tag{1.8}$$

And with the partial derivative chain rule we get the basis vectors for polar coordinates:

$$\frac{\partial}{\partial r} = \frac{\partial x}{\partial r}\frac{\partial}{\partial x} + \frac{\partial y}{\partial r}\frac{\partial}{\partial y} = \cos\theta\frac{\partial}{\partial x} + \sin\theta\frac{\partial}{\partial y},\tag{1.9}$$

$$\text{or}\quad \vec{e}_r = \cos\theta\,\vec{e}_x + \sin\theta\,\vec{e}_y.$$

And,

$$\frac{\partial}{\partial\theta} = \frac{\partial x}{\partial\theta}\frac{\partial}{\partial x} + \frac{\partial y}{\partial\theta}\frac{\partial}{\partial y} = -r\sin\theta\frac{\partial}{\partial x} + r\cos\theta\frac{\partial}{\partial y},\tag{1.10}$$

$$\text{or}\quad \vec{e}_\theta = -r\sin\theta\,\vec{e}_x + r\cos\theta\,\vec{e}_y.$$

We summarize this compactly into a *matrix representation*:

$$\begin{pmatrix}\vec{e}_r & \vec{e}_\theta\end{pmatrix} = \begin{pmatrix}\vec{e}_x & \vec{e}_y\end{pmatrix}\begin{pmatrix}\cos\theta & -r\sin\theta \\ \sin\theta & r\cos\theta\end{pmatrix}.\tag{1.11}$$

Note that subscript entries are grouped together as *row vectors*.

The coordinate basis transformation from the coordinate system $\{x,y\}$ to the coordinate system $\{x',y'\}$ is

$$\begin{pmatrix}\partial_{x'} & \partial_{y'}\end{pmatrix} = \begin{pmatrix}\partial_x & \partial_y\end{pmatrix}\begin{pmatrix}\frac{\partial x}{\partial x'} & \frac{\partial x}{\partial y'} \\ \frac{\partial y}{\partial x'} & \frac{\partial y}{\partial y'}\end{pmatrix}.\tag{1.12}$$

And in general terms, a coordinate basis transformation, *transformation matrix* F, from unprimed coordinates to primed coordinates is

$$\vec{e}_{j'} = \vec{e}_i F^i_{j'}, \quad F^i_{j'} = \frac{\partial x^i}{\partial x^{j'}}; \quad \text{2-dim: } F^i_{j'} = \begin{pmatrix}\frac{\partial x}{\partial x'} & \frac{\partial x}{\partial y'} \\ \frac{\partial y}{\partial x'} & \frac{\partial y}{\partial y'}\end{pmatrix}.\tag{1.13}$$

And what does the reverse transformation look like from primed coordinates to unprimed coordinates? With equation (1.13) we can write:

$$\begin{pmatrix}\partial_x & \partial_y\end{pmatrix} = \begin{pmatrix}\partial_{x'} & \partial_{y'}\end{pmatrix}\begin{pmatrix}\frac{\partial x'}{\partial x} & \frac{\partial x'}{\partial y} \\ \frac{\partial y'}{\partial x} & \frac{\partial y'}{\partial y}\end{pmatrix}.\tag{1.14}$$

or

$$\vec{e}_i = \vec{e}_{k'} R^{k'}_i, \quad R^{k'}_i = \frac{\partial x^{k'}}{\partial x^i}; \quad \text{2-dim: } R^{k'}_i = \begin{pmatrix}\frac{\partial x'}{\partial x} & \frac{\partial x'}{\partial y} \\ \frac{\partial y'}{\partial x} & \frac{\partial y'}{\partial y}\end{pmatrix}.\tag{1.15}$$

Both transformations, (unprimed to primed) F and reverse (primed to unprimed) R, are self evident inverse to each other:

$$R_i^{k'} F_{j'}^i = \begin{pmatrix} \frac{\partial x'}{\partial x} & \frac{\partial x'}{\partial y} \\ \frac{\partial y'}{\partial x} & \frac{\partial y'}{\partial y} \end{pmatrix} \begin{pmatrix} \frac{\partial x}{\partial x'} & \frac{\partial x}{\partial y'} \\ \frac{\partial y}{\partial x'} & \frac{\partial y}{\partial y'} \end{pmatrix}$$

$$= \begin{pmatrix} \frac{\partial x'}{\partial x}\frac{\partial x}{\partial x'} + \frac{\partial x'}{\partial y}\frac{\partial y}{\partial x'} & \frac{\partial x'}{\partial x}\frac{\partial x}{\partial y'} + \frac{\partial x'}{\partial y}\frac{\partial y}{\partial y'} \\ \frac{\partial y'}{\partial x}\frac{\partial x}{\partial x'} + \frac{\partial y'}{\partial y}\frac{\partial y}{\partial x'} & \frac{\partial y'}{\partial x}\frac{\partial x}{\partial y'} + \frac{\partial y'}{\partial y}\frac{\partial y}{\partial y'} \end{pmatrix} \tag{1.16}$$

$$\text{and by chainrule: } = \begin{pmatrix} \frac{\partial x'}{\partial x'} & \frac{\partial x'}{\partial y'} \\ \frac{\partial y'}{\partial x'} & \frac{\partial y'}{\partial y'} \end{pmatrix} = \begin{pmatrix} 1 & 0 \\ 0 & 1 \end{pmatrix}.$$

Or in more general terms:

$$R_i^{k'} F_j^i = \frac{\partial x^{k'}}{\partial x^i}\frac{\partial x^i}{\partial x^{j'}} = \frac{\partial x^i}{\partial x^{j'}}\frac{\partial x^{k'}}{\partial x^i} = F_j^i R_i^{k'} = \delta_{j'}^{k'} = \begin{cases} 1 & j' = k', \\ 0 & j' \neq k'. \end{cases} \tag{1.17}$$

Box 1.3

The transformation matrices $F_{j'}^i$ and $R_i^{k'}$ for polar and spherical coordinates are important formulas. We want to derive them so that we always have them available for later applications.

Polar coordinates:

$$x = r\cos\theta, \quad y = r\sin\theta,$$

$$F_{j'}^i = \begin{pmatrix} F_r^x & F_\theta^x \\ F_r^y & F_\theta^y \end{pmatrix} = \begin{pmatrix} \frac{\partial x}{\partial r} & \frac{\partial x}{\partial \theta} \\ \frac{\partial y}{\partial r} & \frac{\partial y}{\partial \theta} \end{pmatrix} = \begin{pmatrix} \cos\theta & -r\sin\theta \\ \sin\theta & r\cos\theta \end{pmatrix},$$

$$r = \sqrt{x^2 + y^2}, \quad \theta = \tan^{-1}\left(\frac{x}{y}\right), \tag{1.18}$$

$$R_k^{j'} = \begin{pmatrix} R_x^r & R_y^r \\ F_x^\theta & F_y^\theta \end{pmatrix} = \begin{pmatrix} \frac{\partial r}{\partial x} & \frac{\partial r}{\partial y} \\ \frac{\partial \theta}{\partial x} & \frac{\partial \theta}{\partial y} \end{pmatrix} = \begin{pmatrix} \cos\theta & \sin\theta \\ -\frac{\sin\theta}{r} & \frac{\cos\theta}{r} \end{pmatrix}.$$

Spherical coordinates:

$$x = r\sin\theta\cos\phi, \quad y = r\sin\theta\sin\phi, \quad z = r\cos\theta,$$

$$F^i_{j'} = \begin{pmatrix} \frac{\partial x}{\partial r} & \frac{\partial x}{\partial \theta} & \frac{\partial x}{\partial \phi} \\ \frac{\partial y}{\partial r} & \frac{\partial y}{\partial \theta} & \frac{\partial y}{\partial \phi} \\ \frac{\partial z}{\partial r} & \frac{\partial z}{\partial \theta} & \frac{\partial z}{\partial \phi} \end{pmatrix} = \begin{pmatrix} \sin\theta\cos\phi & r\cos\theta\cos\phi & -r\sin\theta\sin\phi \\ \sin\theta\sin\phi & r\cos\theta\sin\phi & r\sin\theta\cos\phi \\ \cos\theta & -r\sin\theta & 0 \end{pmatrix},$$

$$R^{j'}_k = (F^i_{j'})^{-1} = \begin{pmatrix} \sin\theta\cos\phi & \sin\theta\sin\phi & \cos\theta \\ \frac{\cos\theta\cos\phi}{r} & \frac{\cos\theta\sin\phi}{r} & -\frac{\sin\theta}{r} \\ \frac{\sin\phi}{r\cos\theta} & \frac{\cos\phi}{r\cos\theta} & 0 \end{pmatrix}.$$

$$(1.19)$$

Remark 1.5

For index-related quantities such as components, matrices or other multi-index tensors, the order of the factors is irrelevant, as they are all scalars. However, if one has e.g. an arithmetic problem with 'matrix $\times$ matrix', or 'matrix $\times$ vector', so the arrangement of the factors is crucial (e.g. with calculation programmes such as 'mathematica'®). If you hold the rule "subscript index is left and superscript index is right" when summing over an index (Einstein summation), then you are on the safe side. See also Example 1.1.

For multi-index tensors this rule can not always be followed, but it is important to remember the meaning of superscript (contravariant transformation) and subscript (covariant transformation) indices.

∎

Example 1.1

Two matrices A and B are given in a formula context in the following configuration $B^j_k A^i_j$. So this mathematical expression requires a summation over the index j. After the above mentioned, the matrix order has to be exchanged in case of numerical calculation: $A^i_j B^j_k$. In the original order, a computer calculation program would sum (contract) over the indices k and i.

Let the matrix A be $A_j^i = \begin{pmatrix} a & b \\ c & d \end{pmatrix}$ and matrix B be $B_k^j = \begin{pmatrix} 1 & 2 \\ 3 & 4 \end{pmatrix}$. Then $B_k^j A_j^i$ leads to an incorrect result. Which is:

$$B_k^j A_j^i = \begin{pmatrix} 1 & 2 \\ 3 & 4 \end{pmatrix} \begin{pmatrix} a & b \\ c & d \end{pmatrix} = \begin{pmatrix} a + 2c & b + 2d \\ 3a + 4c & 3b + 4d \end{pmatrix}.$$

$A_j^i B_k^j$ leads to a correct result:

$$A_j^i B_k^j = \begin{pmatrix} a & b \\ c & d \end{pmatrix} \begin{pmatrix} 1 & 2 \\ 3 & 4 \end{pmatrix} = \begin{pmatrix} a + 3b & 2a + 4b \\ c + 3d & 2c + 4d \end{pmatrix}.$$

■

A vector $\vec{v}$ in a vector space $\mathcal{V}$ is a defined object which is independent of the selected coordinate system (see Chapter 3, Tensors). The laws of physics exist *idependently* of any coordinate system. A coordinate system is a man made concept, and is introduced to be able to solve a physical problem by calculation. In doing so, one fits the coordinate system to the problem (e.g., polar coordinates in radial symmetry). Therefore:

$$\vec{v} = \vec{e}_i v^i = \vec{e}_{j'} v^{j'}. \tag{1.20}$$

For $\vec{e}_{j'}$, we use the transformation equation (1.13), and get:

$$\vec{v} = \vec{e}_i v^i = \vec{e}_{j'} v^{j'} = (\vec{e}_i F_{j'}^i) v^{j'} = \vec{e}_i F_{j'}^i v^{j'}. \tag{1.21}$$

The transformation relationship between the components is therefore:

$$v^i = F_{j'}^i v^{j'},$$
$$v^{j'} = (F_{j'}^i)^{-1} v^i = R_i^{j'} v^i. \tag{1.22}$$

Comparing this result with (1.13) shows that the transformation behaviour of the basis vectors and the components of a vector is inverse. One defines the transformation of basis vectors (subscript index) as *covariant*, and that of the components of a vector (superscript index) as *contravariant*.

Summary 1.1

$$
\vec{e}_{j'} = \vec{e}_i \, F^i_{j'}, \quad \vec{e}_i = \vec{e}_{j'} \, R^{j'}_i,
$$

$$
v^{j'} = R^{j'}_i \, v^j, \quad v^i = F^i_{j'} v^{j'},
$$

$$
F^i_{j'} = \frac{\partial x^i}{\partial x^{j'}}, \quad R^{j'}_k = \frac{\partial x^{j'}}{\partial x^k}, \tag{1.23}
$$

$$
F^i_{j'} R^{j'}_k = \delta^i_k = \begin{cases} 1 & i = k, \\ 0 & i \neq k. \end{cases}
$$

Chapter 2

Dual Space and Covectors

2.1   Dual Space

Besides the vector space $\mathcal{V}$ there is a complementary space, *the dual vector space* $\mathcal{V}^\star$, or dual space for short. His objects are called *covectors, dual vectors*, or *one-forms* And to distinguish them from the vectors $\vec{v}$ of the vector space $\mathcal{V}$ they are not provided with an arrow but with a tilde $\tilde{p}$. For the dual space, the same vector space axioms from Section 1.1 apply.

The covectors $\tilde{p} \in \mathcal{V}^\star$ have the property that when they "hit" a vector $\vec{v}$ of vector space $\mathcal{V}$, the "product" $\tilde{p}(\vec{v})$ collapses into a number. A similar process is known from quantum mechanics, when a wave function collaps to a numerical value during the measuring process: From a complex, infinitely spreading wave, becomes a sober real number! We hereby conclude: A covector is a *functional* with respect to a vector as an argument:

$$\tilde{p}(\vec{v}) \to \mathbb{R}. \tag{2.1}$$

A covector is also called a *one-form*, 1-form, because it can process a single vector.

Remark 2.1
Covector and *one-form* are two terms that designate the same thing. The term 'covector' is used to illustrate the duality of vector (vector space) and covector (dual space). We speak of a one-form to emphasize the functional processing of a vector into a number.

$\blacksquare$

As already mentioned, the vector space axioms also apply to the covectors. We assume that $\tilde{p}$ and $\tilde{q}$ are two one-forms of $\mathcal{V}^*$, and the following applies:

$$\tilde{s} = \tilde{p} + \tilde{q} \in \mathcal{V}^*,$$
$$\tilde{r} = a\,\tilde{p} \in \mathcal{V}^*, a \in \mathbb{R}. \tag{2.2}$$

One-forms are a linear mapping with respect to vectors, and one thus defines:

$$\tilde{s}(\vec{v}) = (\tilde{p} + \tilde{q})(\vec{v}) = \tilde{p}(\vec{v}) + \tilde{q}(\vec{v}),$$
$$\tilde{r}(\vec{v}) = (a\,\tilde{p})(\vec{v}) = a\,\tilde{p}(\vec{v}), \tag{2.3}$$
$$\tilde{p}(a\,\vec{v} + b\,\vec{w}) = a\,\tilde{p}(\vec{v}) + b\,\tilde{p}(\vec{w}).$$

This mapping (one-form/vector $\to \mathbb{R}$ and vector/one-form $\to \mathbb{R}$) thus forms a vector space.

2.2   Covectors (one-forms)

Covectors (one-forms) of the dual space $\mathcal{V}^*$ are represented, like their "brothers and sisters", from the vector space $\mathcal{V}$ with components and basis vectors:

$$\tilde{p} = p_i\tilde{e}^i. \tag{2.4}$$

Note that a) the *components* have a *subscript* index and b) the *basis* vectors have a *superscript* index. Straight way reverse as with the vectors. The basis vectors, like the covector itself, are provided with a tilde. The literature sometimes the notation $\tilde{d}x^i$ is used for the basis vectors of the dual-space. We will only use this term in a later part of this textbook.

What happens if a one-form basis vector $\tilde{e}^k(\ )$ processes a basis vector $\vec{e}_i$ of the vector space $\mathcal{V}$ as an argument? One defines:

$$\tilde{e}^k(\vec{e}_i) = \delta_i^k = \begin{cases} 1 & i = k, \\ 0 & i \neq k. \end{cases} \tag{2.5}$$

But this process from equation (2.5) is also valid in opposite direction:

$$\vec{e}_k(\tilde{e}^i) = \delta_k^i = \begin{cases} 1 & i = k, \\ 0 & i \neq k. \end{cases} \tag{2.6}$$

From this relationship, the *complementary behavior* of vector space $\mathcal{V}$ and dual space $\mathcal{V}^\star$ becomes clear.

We will now investigate if a covector $\tilde{p}(\ )$ acts on a basis vektor $\vec{e}_i$:

$$\tilde{p}(\vec{e}_i) = (p_k \tilde{e}^k)(\vec{e}_i) = p_k \tilde{e}^k(\vec{e}_i) = p_k \delta_i^k = p_i. \tag{2.7}$$

We conclude that the i-th component of a covector can be selected by means of the i-th basis vector of the vector space. And what happens if the k-th basis vector $\tilde{e}^k(\ )$ of the dual space $\mathcal{V}^\star$ processes a vector $\vec{v}$ of the vector space $\mathcal{V}$ as an argument?

$$\tilde{e}^k(\vec{v}) = \tilde{e}^k(\vec{e}_i v^i) = \tilde{e}^k(\vec{e}_i)v^i = \delta_i^k v^i = v^k. \tag{2.8}$$

And again: If the k-th basis vector of the dual space processes a vector as an argument, it selects the k-th component of the vector.

2.3 Transformation

Using equations (1.23),(2.5) and (2.7), we can derive the following relationship:

$$\tilde{p}(\vec{e}_{j'}) = (p_k \tilde{e}^k)(\vec{e}_i F^i_{j'}) = p_k F^i_{j'} \tilde{e}^k(\vec{e}_i) = p_k F^i_{j'} \delta_i^k = p_k F^k_{j'}. \tag{2.9}$$

LHS: $\tilde{p}(\vec{e}_{j'}) = p_{j'}$. RHS: dummy index $k \to i$. And we obtain:

$$p_{j'} = p_i F^i_{j'}. \tag{2.10}$$

The *components* of a covector transform like basis vectors of $\mathcal{V}$, and thus *covariantly*.

A covector, as well as a vector, is *invariant* under a coordinate transformation: $\tilde{p} = p_i \tilde{e}^i = p_{j'} \tilde{e}^{j'}$. Using equation (2.10), we can derive the following transformation relation for the covector basis:

$$\begin{aligned}
\tilde{p} &= p_i \tilde{e}^i = p_{j'} \tilde{e}^{j'} = p_i F^i_{j'} \tilde{e}^{j'}, \\
\Rightarrow \tilde{e}^i &= F^i_{j'} \tilde{e}^{j'}, \\
\Rightarrow \tilde{e}^{j'} &= (F^i_{j'})^{-1} \tilde{e}^i = R^{j'}_i \tilde{e}^i. \tag{2.11}
\end{aligned}$$

And thus we recognize that the *basis vectors of the dual space* transform like the components of a vector, namely *contravariant*.

Summary 2.1

$$\tilde{e}^{j'} = R_i^{j'}\,\tilde{e}^i, \quad \tilde{e}^i = F_{j'}^i\,\tilde{e}^{j'},$$

$$p_{j'} = p_i F_{j'}^i, \quad p_i = p_{j'} R_i^{j'},$$

$$F_j^i = \frac{\partial x^i}{\partial x^{j'}}, \quad R_k^{j'} = \frac{\partial x^{j'}}{\partial x^k},$$

$$F_{j'}^i R_k^{j'} = \delta_k^i = \begin{cases} 1 & i = k, \\ 0 & i \neq k. \end{cases}$$

$$(2.12)$$

Remark 2.2

When the coordinate system is changed (mapping, transformation, rotation, etc), the "pre-image" is generally to the right of the equals sign (RHS) and the "image" to the left (LHS).

■

2.4   Pictorial representation of covectors (one-forms)

We know the common picture of a vector: an arrow whose length reflects the magnitude of the vector, and the arrowhead the direction. The pictorial representation of a covector (one-form) has nothing in common with an arrow. One has to imagine a gradient field whose *lines have a <u>constant</u> value*, and the direction goes to bigger values. We will learn more about a gradient field in chapter 5.

Figure 2.1 shows an example of a covector (one-form) $\tilde{p} = \tilde{e}^x$. The line grid of the one-form $\tilde{p} = \tilde{e}^x$ expands to infinity. So you can move it in the positive direction, as well as in the negative, without changing its functional character. Therefore, it does not have to coincide with the y-axis. In a 2-dimensional vector space, the covectors (one-forms) are stacks of *curves* of constant values, in a 3-dimensional vector space stacks of *surfaces* of constant values, etc.

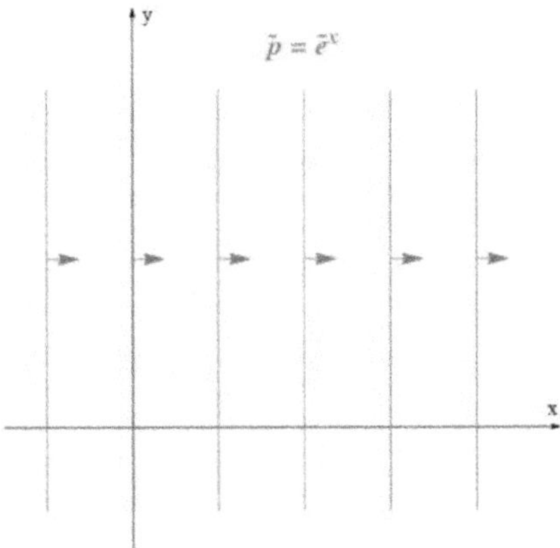

Figure 2.1: Picture of one-form $\tilde{p} = \tilde{e}^x$.

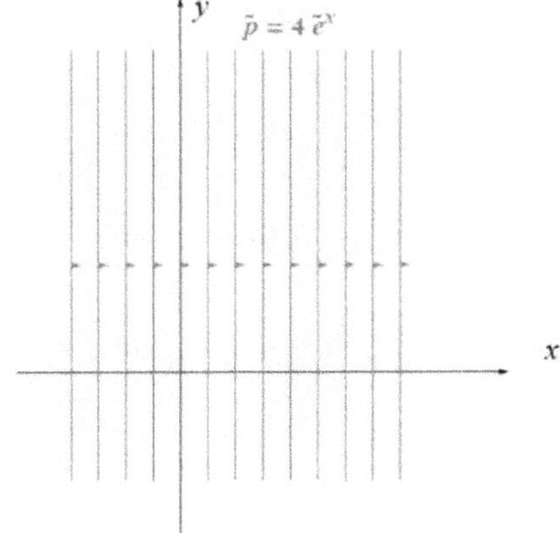

Figure 2.2: Picture of one-form $\tilde{p} = 4\tilde{e}^x$.

The larger the components of a covector, the tighter the lines are stacked. Figure 2.2 shows the picture of covector $\tilde{p} = 4\tilde{e}^x$. How do we visualize the processing of a vector by a one-form to a real number? We apply an arbitrary vector $\vec{v}$ to the line grid of the covector, and count the puncture points of the vector arrow $\vec{v}$ on the covector grid $\tilde{p}$, and thus obtain the real number $\tilde{p}(\vec{v}) \to \mathbb{R}$.

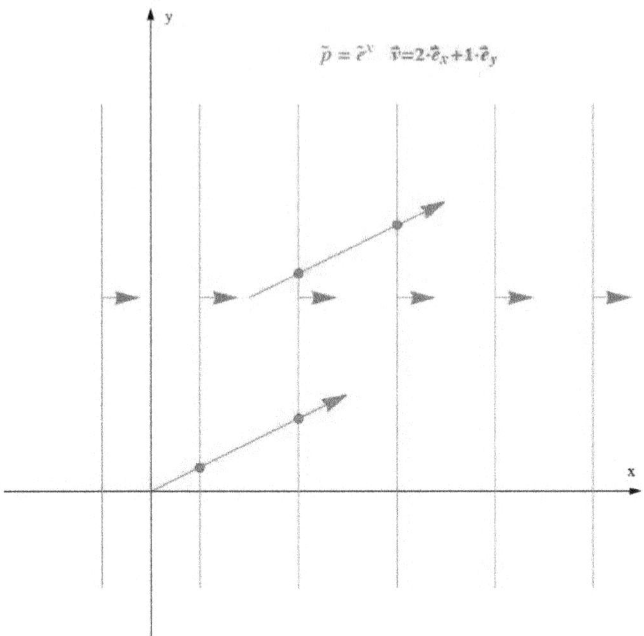

Figure 2.3: The one-form $\tilde{p}$ processes the vector $\vec{v}$ to the number 2.

The Figure 2.3 shows this with the simple example:

$$\tilde{p} = \tilde{e}^x, \quad \vec{v} = 2\,\vec{e}_x + \vec{e}_y,$$
$$\tilde{p}(\vec{v}) = (1 \cdot \tilde{e}^x)(2 \cdot \vec{e}_x + 1 \cdot \vec{e}_y) = 2 \cdot \tilde{e}^x(\vec{e}_x) + 1 \cdot \tilde{e}^x(\vec{e}_y) = 2 \cdot 1 + 1 \cdot 0 = 2.$$

Example 2.1
We want to show a graphic solution of the processing of a vector $\vec{v}$ by a one-form $\tilde{p}$.

$$\vec{v} = 2\,\vec{e}_x + \vec{e}_y,$$
$$\tilde{p} = \tilde{e}^x + 5\,\tilde{e}^y,$$
$$\tilde{p}(\vec{v}) = (\tilde{e}^x + 5\,\tilde{e}^y)(2\,\vec{e}_x + \vec{e}_y) = 1 \cdot 2 + 5 \cdot 1 = 7.$$

To do this in the *same* coordinate system (one-forms and vectors) the values of the one-form components must be mapped with their reciprocal. The graphic solution shows the following Figure 2.4. One can very well recognize in this figure the rule of the linear mapping of one-forms (see also equation (2.3)):

$$\tilde{p}(\vec{v}) = \tilde{p}(\vec{v}_x + \vec{v}_y) = \tilde{p}(\vec{v}_x) + \tilde{p}(\vec{v}_y).$$

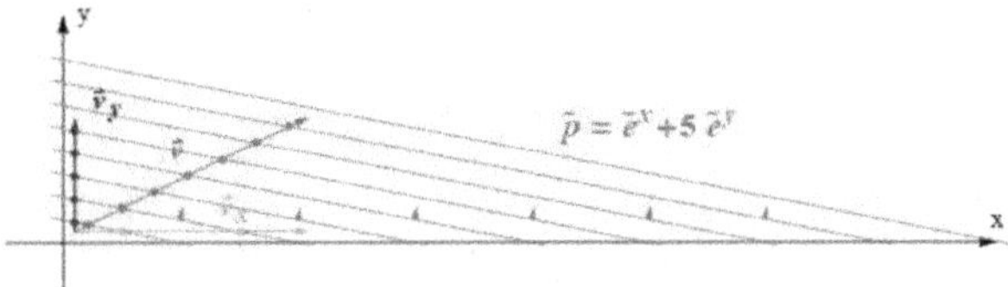

Figure 2.4: Graphic solution of $\tilde{p}(\vec{v}) = (\tilde{e}^x + 5\,\tilde{e}^y)(2\,\vec{e}_x + \vec{e}_y)$.

Chapter 3

Tensors

3.1 Definition and Notation

We start with an abstract definition and then concretize it with a simple first example.

Definition 3.1
A tensor is a *multilinear function* that maps one-forms and vectors as arguments to real numbers, which is linear in each of its arguments (one-forms and vectors).The properties (e.g. symmetry behaviour) of a tensor are not influenced by the selected coordinate system (*frame invariance*). This also applies to tensor equations. Figure 3.1 is a symbolic image of this feature.

■

In Chapter 2 we already learned about a tensor, namely the *one-form*:

1. A one-form maps a vector to a real number:

$$\tilde{p}(\vec{v}) = (p_i \tilde{e}^i)(\vec{e}_j v^j) = p_i v^j \tilde{e}^i(\vec{e}_j) = p_i v^j \delta^i_j = p_i v^i \in \mathbb{R}. \qquad (3.1)$$

$p_i v^i$ is a typical example for the tensor operation: *contraction*. Identical upper and lower indices (within a tensor, or separately in the product of two tensors) are summed up and yield a real number. This operation is *frame invariant*.

2. A one-form is linear in its vector arguments (see equation (2.3)):

$$\tilde{p}(a\,\vec{v} + b\,\vec{w}) = a\,\tilde{p}(\vec{v}) + b\,\tilde{p}(\vec{w}),$$

$$\text{LHS:}\quad p_i\tilde{e}^i(a\,\vec{e}_j v^j + b\,\vec{e}_j w^j) = p_i[(a\,v^j + b\,w^j)\,\underbrace{\tilde{e}^i(\vec{e}_j)}_{\delta^i_j}] = a\,p_i v^i + b\,p_i w^i,$$

$$\text{RHS:}\quad a\,p_i\tilde{e}^i(\vec{e}_j v^j) + b\,p_i\tilde{e}^i(\vec{e}_k w^k) = a\,p_i v^j\,\underbrace{\tilde{e}^i(\vec{e}_j)}_{\delta^i_j} + b\,p_i w^k\,\underbrace{\tilde{e}^i(\vec{e}_j)}_{\delta^i_j} = a\,p_i v^i + b\,p_i w$$

$$(3.2)$$

3. A one-form is frame invariant:

$$\tilde{p} = p_i\tilde{e}^i = \tilde{p}_{j'}\tilde{e}^{j'}. \tag{3.3}$$

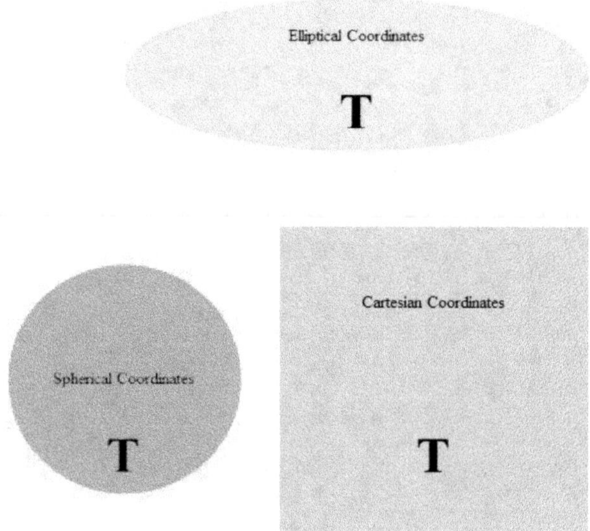

Figure 3.1: No matter in which coordinate system a tensor **T** is located, its properties remain unaffected.

In the notation of a tensor, it is made clear how many one-forms and vectors can be handled. The number of slots for one-forms and vectors are each marked with a number that defines the *rank* of the tensor. The first number is the number of slots for one-forms, the second for vectors. A (0, 2)-tensor can process two vectors. A (1, 0)-tensor can handle only a single one-form. The notation for a tensor $\mathbf{T}$ is as follows:

$$\mathbf{T} \text{ (slots for one-forms ; slots for vectors)} = \mathbf{T}((\tilde{p}, \tilde{q}, \ldots); (\vec{v}, \vec{w}, \ldots)). \quad (3.4)$$

The slot boxes for holding one-forms and vectors are separated by a semicolon. For example, if a tensor only processes vectors, the semicolon may be omitted. The individual slots have their specific meaning, and care must be taken in which slot a vector, a one-form is processed. In general:

$$\mathbf{T}(; (\vec{v}, \vec{w}, \ldots)) \neq \mathbf{T}(; (\vec{w}, \vec{v}, \ldots)) \quad (3.5)$$

However, if the above (0, 2)-tensor is *symmetric* in its vector slots then:

$$\mathbf{T}(; (\vec{v}, \vec{w}, \ldots)) = \mathbf{T}(; (\vec{w}, \vec{v}, \ldots)) \quad (3.6)$$

Multi linearity of a tensor in its arguments signifies, for example, for a (1,1)-tensor:

$$\mathbf{T}(a\,\tilde{p} + b\,\tilde{q}; c\,\vec{v} + d\,\vec{w}) = ac\,\mathbf{T}(\tilde{p}; \vec{v}) + bc\,\mathbf{T}(\tilde{q}; \vec{v}) + ad\,\mathbf{T}(\tilde{p}; \vec{w}) + bd\,\mathbf{T}(\tilde{q}; \vec{w}). \quad (3.7)$$

3.2 Tensor product

A *tensor product* between two objects, denoted by $\otimes$, increases the dimensionality of the representation, in contrast to the usual multiplication from calculus. The tensor product between two covariant basis vectors is represented as follows: $\vec{e}_i \otimes \vec{e}_j$. Without limiting the mathematical accuracy, we will use the shorthand notation $\vec{e}_i \otimes \vec{e}_j = \vec{e}_i \vec{e}_j$.

A short example: Let $\{\vec{e}_i\}$ be the basis of a **2**-dimensional vector space $\mathcal{V}$ $(i = 1, 2)$. The tensor product from these basis vectors

$$\vec{e}_i \vec{e}_j, i, j = 1, 2$$

yields the four basis vectors

$$\vec{e}_1 \vec{e}_1, \ \vec{e}_2 \vec{e}_1, \ \vec{e}_1 \vec{e}_2, \ \vec{e}_2 \vec{e}_2$$

of a **4**-dimensional vector space $\mathcal{V} \otimes \mathcal{V}$.

A basis for any (r, s) tensor can thus be constructed from the tensor product of the vector space basis $\{\vec{e}_i\}$ and the dual space basis $\{\tilde{e}^j\}$:

$$\vec{e}_{i_1} \otimes \vec{e}_{i_2} \otimes \cdots \vec{e}_{i_r} \otimes \tilde{e}^{j_1} \otimes \tilde{e}^{j_2} \otimes \cdots \otimes \tilde{e}^{j_s}. \tag{3.8}$$

We can now express any Tensor $\mathbf{T}$ in component notation as follows

$$\mathbf{T}(\ ;\ ) = T^{ij\cdots}{}_{mn\ldots}\,\vec{e}_i \otimes \vec{e}_j \otimes \cdots \tilde{e}^m \otimes \tilde{e}^n \otimes \cdots (\ ;\ ) = T^{ij\cdots}{}_{mn\ldots}\,\vec{e}_i\vec{e}_j \cdots \tilde{e}^m\tilde{e}^n \cdots (\ ;\ ) \tag{3.9}$$

Remark 3.1
The arrangement of the component indices (superscript, subscript) and their order express the characteristics of a tensor and of course may not be changed!

∎

The components of a tensor are determined by the action of the corresponding basis vectors. For a $(2,1)$-tensor this applies:

$$\begin{aligned}
\mathbf{T}(\tilde{e}^m, \tilde{e}^n; \vec{e}_l) &= T^{ij}{}_k\vec{e}_i\vec{e}_j\tilde{e}^k(\tilde{e}^m, \tilde{e}^n; \vec{e}_l) = T^{ij}{}_k\vec{e}_i(\tilde{e}^m)\vec{e}_j(\tilde{e}^n)\tilde{e}^k(\vec{e}_l) \\
&= T^{ij}{}_k\delta^m_i\delta^n_j\delta^k_l = T^{mn}{}_l.
\end{aligned} \tag{3.10}$$

Each up-index letter represents a one-form (covector) slot, and each lower-index represents a vector slot. For instance, $T^{ij}{}_m$ can thus process two covectors and one vector:

$$\begin{aligned}
\mathbf{T}(\tilde{p}, \tilde{q}; \vec{v}) &= T^{ij}{}_m\vec{e}_i\vec{e}_j\tilde{e}^m(\tilde{p}, \tilde{q}; \vec{v}) = T^{ij}{}_m\vec{e}_i\vec{e}_j\tilde{e}^m(p_r\tilde{e}^r\,q_s\tilde{e}^s\,v^t\vec{e}_t) \\
&= T^{ij}{}_m\,p_r\,\underbrace{\vec{e}_i(\tilde{e}^r)}_{\delta^r_i}\,q_s\,\underbrace{\vec{e}_j(\tilde{e}^s)}_{\delta^s_j}\,v^t\,\underbrace{\tilde{e}^m(\vec{e}_t)}_{\delta^m_t} \\
&= T^{ij}{}_m\,p_iq_jv^m \to \mathbb{R}.
\end{aligned} \tag{3.11}$$

A tensor product generally does not commutate, as the following example shows. Let $\mathbf{R}$ and $\mathbf{S}$ be two $(1,0)$-tensors,

$$\mathbf{R} \otimes \mathbf{S} = R^i\vec{e}_i \otimes S^j\vec{e}_j = R^iS^j\vec{e}_i\vec{e}_j.$$

For example, the tensor component to the basis $\vec{e}_1\vec{e}_2$ of this tensor product is R^1S^2. The commutated tensor product is:

$$\mathbf{S} \otimes \mathbf{R} = S^i\vec{e}_i \otimes R^j\vec{e}_j = S^iR^j\vec{e}_i\vec{e}_j.$$

But now the basis $\vec{e}_1\vec{e}_2$-component is: S^1R^2. And in general: $R^1S^2 \neq S^1R^2$.

Box 3.1

A tensor product always increases the number of free parameters. As an example we take the tensor product of two vectors (column vectors), i.e. two (1,0)-tensors.

$$\vec{v} \otimes \vec{w} = (\vec{e}_i v^i) \otimes (\vec{e}_j v^j) = \vec{e}_i \vec{e}_j v^i w^j = \vec{e}_i \vec{e}_j D^{ij}.$$

And for $i, j = 1, 2$ we obtain:

$$\begin{pmatrix} v^1 \\ v^2 \end{pmatrix} \otimes \begin{pmatrix} w^1 \\ w^2 \end{pmatrix} = \begin{pmatrix} \begin{pmatrix} v^1 \\ v^2 \end{pmatrix} w^1 \\ \begin{pmatrix} v^1 \\ v^2 \end{pmatrix} w^2 \end{pmatrix} = \begin{pmatrix} v^1 w^1 \\ v^2 w^1 \\ v^1 w^2 \\ v^2 w^2 \end{pmatrix} = \begin{pmatrix} D^{11} \\ D^{21} \\ D^{12} \\ D^{22} \end{pmatrix} = D^{ij}.$$

And therefore

$$(1,0) - \text{tensor} \otimes (1,0) - \text{tensor} \rightarrow (2,0) - \text{tensor}.$$

The general expression of a tensor product of an (m, n)-tensor $\mathbf{R}$ and an (r, s)-tensor $\mathbf{S}$ is as follows:

$$\mathbf{R} \otimes \mathbf{S} =$$
$$= R^{i_1 \cdots i_m}_{j_1 \cdots j_n}(\tilde{p}^1, \cdots, \tilde{p}^m; \vec{v}^1, \cdots, \vec{v}^n) \otimes S^{i_1 \cdots i_r}_{j_1 \cdots j_s}(\tilde{p}^{m+1}, \cdots, \tilde{p}^{m+r}; \vec{v}^{n+1}, \cdots, \vec{v}^{n+s})$$
$$= T^{i_1 \cdots i_m i_{m+1} \cdots i_{m+r}}_{j_1 \cdots j_n j_{n+1} \cdots j_{n+s}}(\tilde{p}^1, \cdots, \tilde{p}^m, \tilde{p}^{m+1}, \cdots \tilde{p}^{m+r}; \vec{v}^1, \cdots, \vec{v}^n, \vec{v}^{n+1}, \cdots, \vec{v}^{n+s}).$$

$$(3.12)$$

Note that $\tilde{p}^i$ and $\vec{v}^j$ are different one-forms and vectors. $\mathbf{T}$ is an $(m+r, n+s)$ tensor.

3.3 Tensor Transformation

As we already know, a tensor processes one-forms and vectors into real numbers. Its transformation behaviour when changing coordinates, can thus be derived from the transformation behaviour of the one-forms and vectors. We now want to derive the formula for the transformation of the following tensor

$$\mathbf{T} = T^i_j \, \vec{e}_i \, \tilde{e}^j. \tag{3.13}$$

Remark 3.2
For the term matrix we should note the following: Usually we know the term matrix from linear algebra. There, a matrix processes a vector (column vector) back into a vector (column vector) or a one-form (covector, row vector) into a one-form, as shown for example in equation (3.13). Now, in the tensor algebra, we extend the term matrix to include the components of a rank 2 tensor. For example, the matrix T^{ij} contains the components of a $(2,0)$-tensor. But this matrix can only handle one-forms, but no vectors.

■

By Definition 3.1, a tensor is frame invariant, therefore:

$$\mathbf{T} = T^{i'}_{\ j'}\, \vec{e}_{i'}\, \tilde{e}^{j'} = T^{i}_{\ j}\, \vec{e}_{i}\, \tilde{e}^{j}. \tag{3.14}$$

With equation (1.23) and (2.12) we can transform the LHS of equation (3.14):

$$\text{LHS: } T^{i'}_{\ j'}\, \vec{e}_{i'}\, \tilde{e}^{j'} = T^{i'}_{\ j'}(\vec{e}_{k}F^{k}_{i'})(R^{j'}_{l}\,\tilde{e}^{l}) = \vec{e}_{k}(F^{k}_{i'}T^{i'}_{\ j'}R^{j'}_{l})\tilde{e}^{l} = \vec{e}_{i}(F^{i}_{i'}T^{i'}_{\ j'}R^{j'}_{j})\tilde{e}^{j}. \tag{3.15}$$

In the last step we renamed the dummy indices: $k \to i, l \to j$. We compare the result obtained with the RHS and get:

$$\vec{e}_{i}(F^{i}_{i'}T^{i'}_{\ j'}R^{j'}_{j})\tilde{e}^{j} = T^{i}_{\ j}\, \vec{e}_{i}\, \tilde{e}^{j},$$
$$F^{i}_{i'}T^{i'}_{\ j'}R^{j'}_{j} = T^{i}_{\ j}. \tag{3.16}$$

We multiply from the left with $R^{k'}_{i}$ and from the right with $F^{j}_{l'}$:

$$\underbrace{R^{k'}_{i}F^{i}_{i'}}_{\delta^{k'}_{i'}}\, T^{i'}_{\ j'}\, \underbrace{R^{j'}_{j}F^{j}_{l'}}_{\delta^{j'}_{l'}} = R^{k'}_{i}T^{i}_{\ j}F^{j}_{l'},$$

$$T^{k'}_{\ l'} = R^{k'}_{i}T^{i}_{\ j}F^{j}_{l'} = R^{k'}_{i}F^{j}_{l'}T^{i}_{\ j} = \frac{\partial x^{k'}}{\partial x^{i}}\frac{\partial x^{j}}{\partial x^{l'}}T^{i}_{\ j}. \tag{3.17}$$

The result is that tensors transform like their coordinates. They adapt themselves smoothly to any coordinate system. The tensor transformation formula of equation (3.17) can be extended to any number of indices:

$$T^{i'j'\cdots}_{\ m'n'\ldots} = R^{i'}_{i}R^{j'}_{j}\cdots T^{ij\ldots}_{\ mn\ldots}F^{m}_{m'}F^{n}_{n'}\cdots. \tag{3.18}$$

Chapter 4

Metric Tensors

4.1 Introduction

One of the most important tensors of differential geometry is the metric tensor $\mathbf{g}$. His name already describes one of his 'tasks', namely to determine the magnitude of a vector or a displacement. Just as the geometric space is made up of two spaces (vector space $\mathcal{V}$ and dual space $\mathcal{V}^\star$) that are interwoven, so there is also a second metric tensor, which, among other things, can determine the magnitude of a one-form. We denote it in this text the *dual metric tensor* $\boldsymbol{g}$ because it is capable of processing dual-space one-forms. The metric tensor $\mathbf{g}$ 'lives' in the dual space $\mathcal{V}^\star$, and the dual metric tensor $\boldsymbol{g}$ 'lives' in the vector space $\mathcal{V}$. The Figure 4.1 is an attempt to depict these spatial formations.

The interweaving of vector-space $\mathcal{V}$ and dual space $\mathcal{V}^\star$, of metric tensor $\mathbf{g}$ and dual metric tensor $\boldsymbol{g}$ is further illustrated by the fact that the components of $\mathbf{g} \in \mathcal{V}^\star$ are calculated from the inner product of the basis vectors $\{\vec{e}_i\}$ of the vector space $\mathcal{V}$, Section 4.3.

4.2 Inner product

Up to now we have only created a scalar value from a covector and vector: $\tilde{p}(\vec{v}) \to \mathbb{R}$, or $\vec{v}(\tilde{p}) \to \mathbb{R}$. To produce a scalar value from two vectors a so-called *inner product* is defined: $\vec{v} \cdot \vec{w} \to \mathbb{R}$. The vector space $\mathcal{V}$ in which an inner

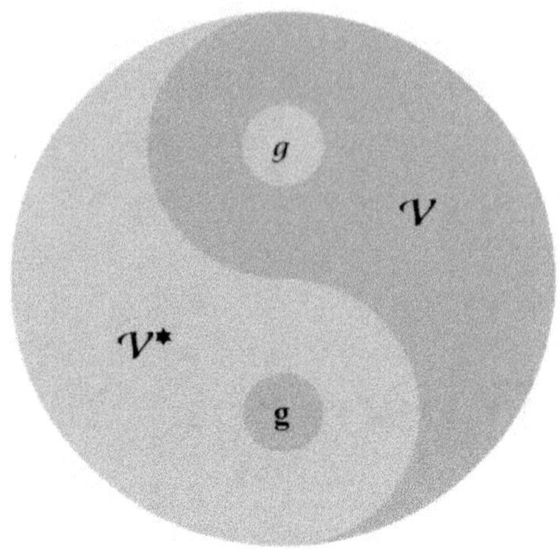

Figure 4.1: A symbol of the interweaving of vector space $\mathcal{V}$ and dual space $\mathcal{V}^\star$ with the metric tensor $\mathbf{g}$ and $\boldsymbol{g}$.

product is defined is called *inner product space*. The following properties must be fulfilled:

1. Bilinearity:

$$
\begin{aligned}
(a\vec{v}_1 + b\vec{v}_2) \cdot \vec{w} &= a(\vec{v}_1 \cdot \vec{w}) + b(\vec{v}_2 \cdot \vec{w}), \\
\vec{v} \cdot (c\vec{w}_1 + d\vec{w}_2) &= c(\vec{v} \cdot \vec{w}_1) + d(\vec{v} \cdot \vec{w}_2)),
\end{aligned}
\tag{4.1}
$$

for all $\vec{v}, \vec{v}_1, \vec{v}_2, \vec{w}, \vec{w}_1, \vec{w}_2 \in \mathcal{V}$ and $a, b, c, d \in \mathbb{R}$.

2. Symmetry:

$$
\vec{v} \cdot \vec{w} = \vec{w} \cdot \vec{v}, \text{ for all } \vec{v}, \ \vec{w} \in \mathcal{V}.
\tag{4.2}
$$

3. Nondegeneracy:

$$
\vec{v} \cdot \vec{x} = 0, \text{ for all } \vec{v} \in \mathcal{V} \Rightarrow \vec{x} = \vec{0}.
\tag{4.3}
$$

We consider in this textbook only symmetrical nondegenerate inner products. For a coordinate basis $\{\vec{e}_i\}$ applies to the inner product:

$$
\vec{e}_i \cdot \vec{e}_j =
\begin{cases}
\mathbb{R} & i = k, \\
0 & i \neq k.
\end{cases}
\tag{4.4}
$$

4.3 Definition and components of g

The *metric tensor* **g** is defined by the inner product of two vectors. A inner product of two vectors yields a real number ($\mathcal{V}$ over $\mathbb{R}$). Therefore, the metric tensor **g** is an $(0, 2)$-tensor, $\mathbf{g}(,) \in \mathcal{V}^\star \otimes \mathcal{V}^\star$, which processes *two vectors as arguments* of any basis into a number. This numerical value is independent of the chosen vector basis. From the above definition follows immediately the linearity of the metric tensor in each of its arguments:

$$\mathbf{g}(a\vec{u} + b\vec{v}, \vec{w}) = (a\vec{u} + b\vec{v}) \cdot \vec{w} = a\vec{u} \cdot \vec{w} + b\vec{v} \cdot \vec{w} = a\,\mathbf{g}(\vec{u}, \vec{w}) + b\,\mathbf{g}(\vec{v}, \vec{w}). \quad (4.5)$$

And in the same way for the second argument. The metric tensor $\mathbf{g}(\vec{v}, \vec{w})$ is is thus a *bilinear form*.

Definition 4.1
The metric tensor **g** is defined by the *inner product* (dot product or scalar product) of two vectors:

$$\mathbf{g}(\vec{v}, \vec{w}) = \vec{v} \cdot \vec{w} = (v^i \vec{e}_i) \cdot (w^j \vec{e}_j) = (\vec{e}_i \cdot \vec{e}_j)\, v^i\, w^j.$$

The component representation of **g** is

$$\mathbf{g}(\,,\,) = g_{ij}\, \tilde{e}^i \otimes \tilde{e}^j(\,,\,) = g_{ij}\, \tilde{e}^i \tilde{e}^j(\,,\,).$$

The components of a tensor are determined by processing basis vectors. And thus we obtain:

$$\begin{aligned}
\mathbf{g}(\vec{e}_m, \vec{e}_n) &= \vec{e}_m \cdot \vec{e}_n, \\
\mathbf{g}(\vec{e}_m, \vec{e}_n) &= g_{ij}\, \tilde{e}^i \tilde{e}^j(\vec{e}_m, \vec{e}_n) = g_{ij}\delta^i_m\delta^j_n = g_{mn}, \\
\Rightarrow g_{mn} &= \vec{e}_m \cdot \vec{e}_n.
\end{aligned} \quad (4.6)$$

For the inner product of two vectors we get:

$$\vec{v} \cdot \vec{w} = g_{ij}\, v^i\, w^j. \quad (4.7)$$

Remark 4.1
Since basic vectors are generally location-dependent in direction and magnitude (polar coordinates, spherical coordinates, etc.), the inner product must always be calculated based on Cartesian coordinates.

■

Example 4.1
We determine the components of the metric tensor for *polar coordinates*. With equation (1.11) we get:

$$
\begin{aligned}
\vec{e}_r \cdot \vec{e}_r &= (\cos\theta\,\vec{e}_x + \sin\theta\,\vec{e}_y) \cdot (\cos\theta\,\vec{e}_x + \sin\theta\,\vec{e}_y) = 1, \\
\vec{e}_r \cdot \vec{e}_\theta &= (\cos\theta\,\vec{e}_x + \sin\theta\,\vec{e}_y) \cdot (-r\sin\theta\,\vec{e}_x + r\cos\theta\,\vec{e}_y) = 0, \\
\vec{e}_\theta \cdot \vec{e}_r &= (-r\sin\theta\,\vec{e}_x + r\cos\theta\,\vec{e}_y) \cdot (\cos\theta\,\vec{e}_x + \sin\theta\,\vec{e}_y) = 0, \\
\vec{e}_\theta \cdot \vec{e}_r &= (-r\sin\theta\,\vec{e}_x + r\cos\theta\,\vec{e}_y) \cdot (-r\sin\theta\,\vec{e}_x + r\cos\theta\,\vec{e}_y) = r^2.
\end{aligned}
$$

The components of the metric tensor for polar coordinates $\{r,\theta\}$ are thus:

$$
g_{rr} = 1, \quad g_{r\theta} = g_{\theta r} = 0, \quad g_{\theta\theta} = r^2. \tag{4.8}
$$

■

Since the inner product is symmetric, $\mathcal{V}$ over $\mathbb{R}$, $\vec{v} \cdot \vec{w} = \vec{w} \cdot \vec{v}$, the metric tensor is also *symmetric in its arguments*. Another argument for the symmetry of the metric tensor is as follows. In Cartesian coordinates:

$$
g_{ij} = \vec{e}_i \cdot \vec{e}_j = \delta_{ij} = \delta_{ji} = g_{ji}. \tag{4.9}
$$

g_{ij} is therefore symmetrical in Cartesian coordinates, and thus symmetrical in all other coordinate systems, because the properties of a tensor are frame invariant (Definition 3.1). Therefore

$$
\mathbf{g}(\vec{v},\vec{w}) = \mathbf{g}(\vec{w},\vec{v}), \quad g_{ij} = g_{ji}. \tag{4.10}
$$

We now analyse the transformation behaviour of the metric components g_{ij}. With equation (4.6) and (1.13) applies

$$
g_{i'j'} = \vec{e}_{i'} \cdot \vec{e}_{j'} = (\vec{e}_i F^i_{i'}) \cdot (\vec{e}_j F^j_{j'}) = (\vec{e}_i \cdot \vec{e}_j) F^i_{i'} F^j_{j'} = g_{ij} F^i_{i'} F^j_{j'}. \tag{4.11}
$$

If we write the (0,2) metric tensor g_{ij} in matrix form $(g) \equiv (g^i_j)$, in order to be able to continue to calculate with matrices, we must keep to the matrix multiplication rules (see Remark 1.5), and adjust the transformation matrix F accordingly. And thus equation (4.11) is written in matrix notation:

$$\left(g^i_{j'}\right) = \left(F^i_{i'}\right)^T \left(g^i_j\right) F^j_{j'},$$
$$\left(g'\right) = F^T(g)\,F.$$
(4.12)

Example 4.2

Using equation (4.12) we want to determine the metric matrix $\left(g'\right)$ for polar coordinates. In Box 1.3 we have already determined the transformation matrix $F = \begin{pmatrix} \cos\theta & -r\sin\theta \\ \sin\theta & r\cos\theta \end{pmatrix}$ for polar coordinates and with $(g) = \delta^i_j$ we get

$$\left(g'\right)_{\text{polar}} = \begin{pmatrix} \cos\theta & -r\sin\theta \\ \sin\theta & r\cos\theta \end{pmatrix}^T \begin{pmatrix} 1 & 0 \\ 0 & 1 \end{pmatrix} \begin{pmatrix} \cos\theta & -r\sin\theta \\ \sin\theta & r\cos\theta \end{pmatrix} = \begin{pmatrix} 1 & 0 \\ 0 & r^2 \end{pmatrix}. \quad (4.13)$$

∎

Definition 4.2

The norm of a vector $\| \vec{v} \|$ is defined by the inner product:

$$\| \vec{v} \|^2 = \vec{v} \cdot \vec{v} = \mathbf{g}(\vec{v}, \vec{v}) = g_{ij} v^i v^j. \quad (4.14)$$

∎

4.4 Signature of the metric

The metric tensor $\mathbf{g}$ is symmetrical in its indices, Section 4.3 and equation (4.10). According to the spectral theorem, any symmetrical matrix can be diagonalized, has real eigenvalues Λ and and orthogonal eigenvectors Q. Thus applies:

$$\left(g'_{\text{diagonal}}\right) = \Lambda = Q^T (g)\,Q. \quad (4.15)$$

You choose Q so that the negative eigenvalues in Λ come first. Then we normalize the eigenvalues with a diagonal matrix D.

$$\left(g'_{\text{canonical}}\right) = D^T \Lambda D = D^T Q^T \left(g\right) QD = (QD)^T \left(g\right) QD =$$
$$= F^T(g)F = \operatorname{diag}(-1,\ldots,-1,1,\ldots,1). \tag{4.16}$$

This is the theorem that every vector space with a symmetrical metric tensor has an *orthonormal basis* by which it can be transformed into a canonical form. The number of +1's and -1's is independent of the selected basis. The number s of minus signs is called *signature* of the metric **g**.

The transformation $\left(g'\right) = F^T(g)\,F$, equation (4.12), is a similarity transformation and thus the canonical form $\left(g'_{\text{canonical}}\right)$ describes the vector space in the same way as the original metric did. If $\left(g'_{\text{canonical}}\right)$ consists of only 1's, we speak of a positively definite *Euclidean flat space* or *Riemannian curved space* . If the first entry is a -1 followed by +1's, this is a *Minkowski flat space* (special relativity with $n = 4$) or *Lorentzian curved space* (general relativity with $n = 4$).

The canonical form of the metric tensor selects the orthonormal basis from the set of bases of a vector space. In Euclidean space such a basis is called *Cartesian*. Because of

$$g_{ij} = \delta_{ij}, \quad (g) = E,$$
$$E = F^T E F = F^T F \;\Rightarrow\; F^T = F^{-1} \;\Rightarrow\; F \text{ orthogonal}, \tag{4.17}$$

orthogonal matrices are the transformation between the Cartesian bases.

The Minkowski metric

$$\eta = \begin{pmatrix} -1 & 0 & 0 & 0 \\ 0 & 1 & 0 & 0 \\ \vdots & \vdots & \ddots & \vdots \\ 0 & 0 & 0 & 1 \end{pmatrix} = \operatorname{diag}(-1,1,\ldots,1) \tag{4.18}$$

selects the so-called *Lorentz* basis. The Lorentz transformation matrix (basis transformation) L fulfils the relation

$$\eta = L^T \eta\, L. \tag{4.19}$$

Note that η is not E and therefore L is not orthogonal.

4.5 Integration

The magnitude of any infinitesimal displacement $d\vec{l} = dx^i \vec{e}_i$ is determined using the metric tensor as follows:

$$ds^2 = d\vec{l} \cdot d\vec{l} = \mathbf{g}\left(d\vec{l}, d\vec{l}\right) = g_{ij}\tilde{e}^i\tilde{e}^j\left(dx^k\vec{e}_k, dx^l\vec{e}_l\right)$$
$$= g_{ij}dx^k dx^l \underbrace{(\tilde{e}^i\vec{e}_k)}_{\delta^i_k}\underbrace{(\tilde{e}^j\vec{e}_l)}_{\delta^j_l} = g_{ij}dx^i dx^j \tag{4.20}$$

$$ds^2 = g_{ij}dx^i dx^j.$$

And with the result from Example 4.1 we get for a line element in polar coordinates:

$$ds^2 = (dr)^2 + r^2(d\theta)^2. \tag{4.21}$$

To determine the length l of a curve the integration is done as follows

$$l = \int_{\text{curve}} ds = \int_{\text{curve}} \left|\sqrt{g_{ij}dx^i dx^j}\right| = \int_{\lambda_0}^{\lambda_1} \left|\sqrt{g_{ij}\frac{dx^i}{d\lambda}\frac{dx^j}{d\lambda}}\right| d\lambda, \tag{4.22}$$

where the curve is parametrized with λ. The tangential vector $\vec{u}$ of the curve has the coordinates $u^i = \frac{dx^i}{d\lambda}$ (sec Section 7.2) and so we finally get

$$l = \int_{\lambda_0}^{\lambda_1} \left|\sqrt{g_{ij}\frac{dx^i}{d\lambda}\frac{dx^j}{d\lambda}}\right| d\lambda = \int_{\lambda_0}^{\lambda_1} \left|\sqrt{\vec{u}\cdot\vec{u}}\right| d\lambda. \tag{4.23}$$

For volume integration ($n = 3$, Euclidean/Riemannian space), as known from calculus, the so-called Jacobian, the determinant of the transformation matrix is used. The following applies to a transformation from $\{x^{i'}\}$ to $\{x^i\}$:

$$\underbrace{dx^1 dx^2 dx^3}_{\text{volume element } x^i \text{ coordinates}} = \underbrace{\frac{\partial\left(x^1, x^2, x^3\right)}{\partial\left(x^{1'}, x^{2'}, x^{3'}\right)} dx^{1'} dx^{2'} dx^{3'}}_{\text{volume element } x^{i'} \text{ coordinates}}. \tag{4.24}$$

The Jacobian J is known to be defined as

$$\frac{\partial\left(x^1, x^2, x^3\right)}{\partial\left(x^{1'}, x^{2'}, x^{3'}\right)} = \det\begin{pmatrix} \frac{\partial x^1}{\partial x^{1'}} & \frac{\partial x^2}{\partial x^{1'}} & \frac{\partial x^3}{\partial x^{1'}} \\ \frac{\partial x^1}{\partial x^{2'}} & \frac{\partial x^2}{\partial x^{2'}} & \frac{\partial x^3}{\partial x^{2'}} \\ \frac{\partial x^1}{\partial x^{3'}} & \frac{\partial x^2}{\partial x^{3'}} & \frac{\partial x^3}{\partial x^{3'}} \end{pmatrix} = \det\left(F^i_{\,j'}\right) = \det(F) = J,$$

$$\tag{4.25}$$

where $F^i_{\;j'}$ is the known transformation matrix. The task in volume integration, equation (4.24), is therefore to determine the Jacobian which can be a very arduous job. A simple way to do this is to use equation (4.12):

$$\left(g'\right) = F^T(g)F,$$
$$\det\left(g'\right) = \det\left(F^T\right)\det(g)\det(F).$$

$$(4.26)$$

In 3-dimensional Cartesian space, $\det(g) = 1$, and with the validity of $\det\left(F^T\right) = \det(F)$ for matrices, the Jacobian J results in:

$$\det(F) = J = \sqrt{\det\left(g'\right)}.$$

$$(4.27)$$

With equation (4.24) we get the following result:

$$dx^1 dx^2 dx^3 = \sqrt{\det\left(g'\right)}\, dx^{1'} dx^{2'} dx^{3'}.$$

$$(4.28)$$

Example 4.3

We want to determine the metric tensor for spherical coordinates $g_{i'j'} = \left(g'\right)$ and then the corresponding Jacobian $\sqrt{\det\left(g'\right)}$. We know the transformation matrix for spherical coordinates from Box 1.3

$$F^i_{\;j'} = \begin{pmatrix} \sin\theta\cos\phi & r\cos\theta\cos\phi & -r\sin\theta\sin\phi \\ \sin\theta\sin\phi & r\cos\theta\sin\phi & r\sin\theta\cos\phi \\ \cos\theta & -r\sin\theta & 0 \end{pmatrix}.$$

And with equation (4.12) we thus obtain the metric tensor for spherical coordinates:

$$g_{i'j'} = \left(F^i_{\;i'}\right)^T g_{ij}\, F^j_{\;j'} = \left(F^i_{\;i'}\right)^T \delta_{ij}\, F^j_{\;j'} = \begin{pmatrix} 1 & 0 & 0 \\ 0 & r^2 & 0 \\ 0 & 0 & r^2\sin^2\theta \end{pmatrix}.$$

And the Jacobian J for volume integration at spherical coordinates is

$$J = \sqrt{\det\left(g'\right)} = \sqrt{\det\begin{pmatrix} 1 & 0 & 0 \\ 0 & r^2 & 0 \\ 0 & 0 & r^2\sin^2\theta \end{pmatrix}} = r^2\sin^2\theta.$$

And the proper volume element in these coordinates

$$r^2\sin\theta\, dr\, d\theta\, d\phi. \qquad \blacksquare$$

The volume integration in the Minkowski space changes to the previously said only by the metric tensor of the orthogonal coordinates:

$$g_{ij} = \begin{cases} -1, & i = j = 0, \\ +1, & i = j = 1,2,3 \\ 0 & i \neq j \end{cases} \qquad \Rightarrow \det(g_{ij}) = -1. \qquad (4.29)$$

According to the equation (4.24) the relation for the Minkowski space is

$$dx^0 dx^1 dx^2 dx^3 = \frac{\partial\left(x^0, x^1, x^2, x^3\right)}{\partial\left(x^{0'}, x^{1'}, x^{2'}, x^{3'}\right)} \, dx^{0'} dx^{1'} dx^{2'} dx^{3'}. \qquad (4.30)$$

And according to the previous explanations we determine the Jacobian J:

$$\det\left(g'\right) = \det\left(F^T\right) \det\left(g\right) \det\left(F\right) = \det\left(F^T\right)(-1)\det\left(F\right),$$

$$\det\left(F\right) = J = \sqrt{-\det\left(g'\right)}. \qquad (4.31)$$

And correspondingly for the volume integration one obtains

$$dx^0 dx^1 dx^2 dx^3 = \sqrt{-\det\left(g'\right)} \, dx^{0'} dx^{1'} dx^{2'} dx^{3'}. \qquad (4.32)$$

4.6 Tensor densities

As we have already mentioned in Section 4.5 and equation (4.24) in order to keep a scalar value, i.e. a tensor of rank zero, (e.g. a volume V, a mass m, a charge q, etc.) **constant** during the coordinate transformation $\{x^{i'}\}$ to $\{x^i\}$ the Jacobian correction term is required in integration.

$$\begin{aligned}
\text{volume integration} \quad & \underbrace{dx^1 dx^2 dx^3}_{dV} = \underbrace{J \, dx^{1'} dx^{2'} dx^{3'}}_{dV'}, \\[2ex]
\text{mass integration} \quad & \underbrace{\rho_m \, dx^1 dx^2 dx^3}_{dm} = \underbrace{\rho_{m'} \, dx^{1'} dx^{2'} dx^{3'}}_{dm'}, \\[2ex]
\text{charge integration} \quad & \underbrace{\rho_q \, dx^1 dx^2 dx^3}_{dq} = \underbrace{\rho_{q'} \, dx^{1'} dx^{2'} dx^{3'}}_{dq'}.
\end{aligned} \qquad (4.33)$$

From the mere volume integration we get: $J dx^{i'} = dx^i \Rightarrow dx^{i'} = J^{-1} dx^i$. And from this we see that with the correction factor J^{-1} a length segment dx^1, an area segment $dx^1 dx^2$, a volume segment $dx^1 dx^2 dx^3$, etc., is transformed like a tensor of rank zero. Because of $dm' = dm$ or $dq' = dq$ we receive:

$$
\begin{aligned}
\rho_{q'} dx^{i'} &= \rho dx^i \\
\rho_{q'} J^{-1} dx^i &= \rho dx^i \\
\rho_{q'} &= J \rho.
\end{aligned}
\tag{4.34}
$$

A density (mass, charge, etc.) has J as correction factor.

Scalar objects $\mathfrak{s}$ that transform with the help of the transformation determinant J are called **tensor densities**:

$$
\mathfrak{s}' = J^w \, \mathfrak{s}.
\tag{4.35}
$$

The power of the Jacobian J is called *weight*. Thus a volume element is a scalar density of weight -1, a charge density (or mass density) is a scalar density of weight +1. If we consider the determinant g of the metric tensor g_{ij}, we can see with the help of the equation (4.26) that it is a tensor density of weight +2:

$$
\det\left(g'\right) = \det\left(F^T\right)\det(g)\det(F) = J^2 \det(g).
\tag{4.36}
$$

Remark 4.2

Some authors define Jacobian as $J = \dfrac{\partial\left(x^{i'}\right)}{\partial(x^i)}$, i.e. a transformation from $\{x^i\}$ to $\{x^{i'}\}$, inverse to our text. Therefore the weights then have different signs.

■

The Levi-Civita symbol $\epsilon_{i_1 i_2 \ldots i_n}$ is a *completely antisymmetric* function of n indices, which have values from 1 to n. It is defined as follows

$$
\epsilon_{i_1 i_2 \ldots i_n} = \epsilon^{i_1 i_2 \ldots i_n} =
\begin{cases}
+1, & i_1 i_2 \ldots i_n \text{ is an even permutation of } 12 \ldots n \\
-1, & i_1 i_2 \ldots i_n \text{ is an odd permutation of } 12 \ldots n \\
0 & i_1 i_2 \ldots i_n \text{ is not a permutation of } 12 \ldots n
\end{cases}
\tag{4.37}
$$

The number of even permutations is $n!/2$, the number of odd permutations is also $n!/2$. In total $\epsilon_{i_1 i_2 \ldots i_n}$ can take n^n values.

Box 4.1

A permutation is a rearrangement of a sequence of integers $(12 \ldots n)$, for example in the order $(i_1 i_2 \ldots i_n)$. There are $n!$ different permutations of n integers.

To feed back any arrangement, e.g. $(i_1 i_2 \ldots i_n)$, to the initial arrangement $(12 \ldots n)$ one can proceed as follows: If $i_1 = 1$, then nothing is to be done. If $i_j = 1$, i_1 is exchanged with i_j. Such an exchange is called *transposition*. This process is repeated until the natural arrangement is reached again. The number of transpositions indicates the *parity* of the permutation $(i_1 i_2 \ldots i_n)$. If the number is an even number, the parity is even. If it is an odd number, the parity is odd.

We want to investigate the transformation behaviour of the Levi-Civita symbol $\epsilon_{i_1 i_2 \ldots i_n}$. In the following we will show that it is a tensor density, with weight $+1$. We use the determinant expansion, see Appendix B.1, equation (B.4),

$$\det(A) = \epsilon_{i_1 i_2 \ldots i_n} a_1^{i_1} a_2^{i_2} \cdots a_n^{i_n}. \tag{4.38}$$

If $\epsilon_{i_1 i_2 \ldots i_n}$ were a $(0, n)$-tensor, the following transformation rule would apply:

$$\epsilon'_{j_1 j_2 \ldots j_n} = \epsilon_{i_1 i_2 \ldots i_n} F_{j_1}^{i_1} F_{j_2}^{i_2} \ldots F_{j_n}^{i_n}. \tag{4.39}$$

If we compare the RHS of equation (4.39) with equation (4.38), we see that it is equal to $(\det F)$ at $j_1 j_2 \ldots j_n$ as an even permutation with respect to $i_1 i_2 \ldots i_n$, and $-(\det F)$ at odd permutation. And thus

$$\epsilon'_{j_1 j_2 \ldots j_n} = (\det F)\, \epsilon_{j_1 j_2 \ldots j_n} = J\, \epsilon_{j_1 j_2 \ldots j_n}. \tag{4.40}$$

Thus the Levi-Civita symbol is transformed according to the tensor transformation rule, but with a weighting factor J. The weight is therefore $+1$.

4.7 Mapping of vectors into one-forms (covectors)

As we learned in Section 4.3, the metric tensor $\mathbf{g}(\ ,\ )$ processes two vectors into a number. But if you give it only one vector for processing $\mathbf{g}(\vec{v},\ )$, then the remaining tensor is a *one-form* $\tilde{v}(\ )$, a functional that maps another vector to a real number. Therefore:

$$\mathbf{g}(\vec{v},\ ) = \tilde{v}(\ ). \tag{4.41}$$

We denote this one-form with the same letter as a vector, because they are *complementary* quantities, as we will see in the following.

$$\begin{aligned} v_i &= \tilde{v}(\vec{e}_i) = \mathbf{g}(\vec{v}, \vec{e}_i) = \vec{v} \cdot \vec{e}_i = \\ &= \vec{e}_i \cdot \vec{v} = \vec{e}_i \cdot (\vec{e}_j) v^j = \vec{e}_i \cdot \vec{e}_j v^j = g_{ij} v^j, \\ v_i &= g_{ij} v^j. \end{aligned} \tag{4.42}$$

Hereby it was shown that vector components can be converted into covector components with the help of the metric tensor. As we will see later, it also works in the other direction, namely through the dual metric tensor (Section 4.8). The metric tensor, is the switchboard that establishes a *complementarity* between vectors and one-forms (covectors), between vector space $\mathcal{V}$ and dual space $\mathcal{V}^\star$.

4.8 Dual metric tensor g

In Section 4.7, we learned that the metric tensor $\mathbf{g}$ reshapes a vector $\vec{v}$ into 'its' one-form $\tilde{v}$. In other words, the metric tensor transforms the upper-index of a vector component into the lower-index of a one-form component, equation (4.42).

Is the process from equation (4.42) also reversible? That can a vector be found which is related to given one-form? The answer can generally be answered immediately with yes, because vector space and dual space, or vector and one-form, are complementary quantities, are the two sides of the same coin, see Figure 4.1. Therefore, the *dual metric tensor* g is a (2, 0)-tensor, $g \in \mathcal{V} \otimes \mathcal{V}$, which processes *two one-forms as arguments* of any basis into a number. If $\boldsymbol{g}$ processes only a single one-form as argument, the result is a vector, the complement to the processed one-form.

$$\begin{aligned} \boldsymbol{g}(\tilde{p}, \tilde{q}) &\to \mathbb{R}, \\ \boldsymbol{g}(\tilde{v}, \ ) &= \vec{v}. \end{aligned} \tag{4.43}$$

We define the dual metric tensor $\boldsymbol{g}$ as follows:

Definition 4.3
Dual metric tensor $\boldsymbol{g}(\, , \,) = \mathrm{g}^{ij}\vec{e}_i \otimes \vec{e}_j = \mathrm{g}^{ij}\vec{e}_i\vec{e}_j \in \mathcal{V} \otimes \mathcal{V}.$

$$\mathrm{g}^{ij}\,g_{jk} = \delta^i_k. \tag{4.44}$$

With this definition it is clear that metric and dual metric are truly inverse to each other. Therefore, from now on we will unify the chosen letter form for *metric components*. With metric tensor components: g_{jk}, and dual metric tensor components: g^{ij}, we get:

$$g^{ij}\,g_{jk} = \delta^i_k. \tag{4.45}$$

$\blacksquare$

4.9 Index raising and lowering

We apply the Definition 4.3 of the dual metric tensor $\boldsymbol{g}$ to equation (4.42) and obtain:

$$\begin{aligned} v_j &= g_{jk}v^k, \\ g^{ij}v_j &= g^{ij}g_{jk}v^k = \delta^i_k v^k = v^i, \\ &\Longrightarrow \\ v^i &= g^{ij}v_j. \end{aligned} \tag{4.46}$$

Using equation (4.42) and equation (4.46) we have created the arithmetic basics to transform a vector $\vec{v}$ (vector space $\mathcal{V}$) into its complementary object, the corresponding one-form $\tilde{v}$ (dual space $\mathcal{V}^\star$), and vice versa. This process is called index lowering ($\vec{v} \longrightarrow \tilde{v}, v^i \longrightarrow v_j$) or index raising ($\tilde{v} \longrightarrow \vec{v}, v_i \longrightarrow v^j$).

> **Summary 4.1**
>
> *index lowering* (replacing a vector component with a one-form component):
>
> $$v^i \longrightarrow v_j, \ v^i = g^{ij} v_j.$$
>
> *index raising* (replacing a one-form component with a vector component):
>
> $$v_i \longrightarrow v^j, \ v_i = g_{ij} v^j.$$

Example 4.4

We want to show a practical example of where the application of index lowering and raising is used. We choose Cartesian coordinates for the vector space and get the following expressions for the metric tensor (Definition 4.1) and for the inner product, equation (4.7):

$$g_{ij} = \delta_{ij}, \quad \vec{v} \cdot \vec{w} = g_{ij} v^i w^j = \delta_{ij} v^i w^j = v_j w^j.$$

If we want to lower the index of w^i we have to use the dual metric tensor g^{ij}:

$$w^i = w_j g^{ij} = w_j \delta^{ij}.$$

The electrostatic potential energy U of two interacting dipoles with the dipole moments $\vec{P}$ and $\vec{Q}$ is given by $U = \frac{1}{r^3} \left\{ \vec{P} \cdot \vec{Q} - \frac{3(\vec{P} \cdot \vec{r})(\vec{Q} \cdot \vec{r})}{r^2} \right\}$, where $\vec{r}$ is the position vector from $\vec{P}$ to $\vec{Q}$. We look for the dipole-coupling tensor $\mathbf{T}(\,;\vec{P}, \vec{Q}) = U$, i.e. a $(0, 2)$-tensor.

For the occurring dot products we can write:

$$\vec{P} \cdot \vec{Q} = P_i Q^i, \ \vec{P} \cdot \vec{r} = P_i r^i, \text{ and } \vec{Q} \cdot \vec{r} = Q_j r^j.$$

The solution strategy is to 'bracket out' the $\vec{P}$ and $\vec{Q}$ terms from the formula. The remaining bracket term is then the coupling tensor between $\vec{P}$ and $\vec{Q}$. As you can see, $\vec{Q}$ has an up-index in the first inner product and a lower-index in the last one. In order to be able to bracket out the $\vec{Q}$-component, we want to lower the index in the first inner product: $P_i Q^i = P_i \delta^{ij} Q_j$. And thus

$$\vec{P} \cdot \vec{Q} - \frac{3(\vec{P} \cdot \vec{r})(\vec{Q} \cdot \vec{r})}{r^2} = P_i \delta^{ij} Q_j - \frac{3}{r^2} P_i x^i Q_j x^j$$

$$= P_i Q_j \left(\delta^{ij} - \frac{3}{r^2} x^i x^j \right)$$

$$= P_i Q_j T^{ij},$$

with

$$T^{ij} = \delta^{ij} - \frac{3}{r^2} x^i x^j.$$

T^{ij}, however, are the components of a $(2,0)$ tensor as opposed to our initial assumption of a $(0,2)$ tensor that has two vectors to process. In a Cartesian coordinate system, as used here, the components of one-forms $\{P_i, Q_j\}$ and vectors $\{P^k, Q^l\}$ are identical:

$$P_i = \delta_{ik} P^k, \quad Q_j = \delta_{jl} Q^l.$$

We transfer the above result

$$P_i Q_j T^{ij} = P^k Q^l \delta_{ik} \delta_{jl} T^{ij} = P^k Q^l T_{kl}, \text{ with } T_{kl} = \delta_{kl} - \frac{3}{r^2} x_k x_l.$$

The electrostatic potential energy U can thus be written very compact as

$$U = \mathbf{T}\left(\,; \vec{P}, \vec{Q}\right) = \frac{1}{r^3} P^k Q^l T_{kl} = \frac{1}{r^3} P^k Q^l \left(\delta_{kl} - \frac{3}{r^2} x_k x_l\right).$$

$\blacksquare$

4.10 Magnitudes and dot products of one-forms

A one-form $\tilde{v}$ is also defined as having the same magnitude as its associated vector $\vec{v}$. Therefore

$$\|\tilde{v}\|^2 = \|\vec{v}\|^2 = g_{ij} v^i v^j. \tag{4.47}$$

With equation (4.46) we are able to convert the vector components into one-form components (index lowering). And so we get:

$$\|\tilde{v}\|^2 = g_{ij} v^i v^j = g_{ij} \left(g^{ik} v_k\right)\left(g^{jl} v_l\right) = v_k v_l\, g^{ik} \left(g_{ij} g^{jl}\right)$$
$$= v_k v_l\, g^{ik} \delta_i^l = v_k v_l\, g^{lk}.$$

And we can write for the *norm of one-form*:

$$\|\tilde{v}\|^2 = v_k v_l\, g^{lk}. \tag{4.48}$$

And so we can also define the inner product of one-forms:

$$\tilde{v} \cdot \tilde{w} = v_k v_l \, g^{lk}. \tag{4.49}$$

4.11 Compilation of the handling of tensors

The *contraction*, the summation of an upper index and a lower index, transforms a (r, s) tensor into a $(r-1, s-1)$ tensor. It is important to note the position of the indices. For the sake of clarity, we will therefore arrange the upper indices and the lower indices staggered.

$$\text{index position number}: T^{(1)(2)\cdots(r)}{}_{(r+1)(r+2)\cdots(r+s)}, \; T^{(1)}{}_{(2)}{}^{(3)}, \; T_{(1)(2)}{}^{(3)}, \ldots;$$

$$(2,4) - \text{contraction}: T^{ij}{}_{kjl}, \; j - \text{up position 2}, \; j - \text{low position 4};$$

$$T^{ij}{}_{kjl} \neq T^{ij}{}_{jkl} : (2,4) - \text{contraction} \neq (2,3) - \text{contraction}.$$

As we have already mentioned, indices that are not summed up are called *free* indices, and those that are summed up (contracted) are called *dummy* indices.

The metric is used to *raise* or *lower* indices . Starting from a tensor $T^{ij}{}_{kl}$, we can define new tensors:

$$T^{ijm}{}_{l} = g^{mk} \, T^{ij}{}_{kl},$$
$$T^{ij}{}_{k}{}^{m} = g^{ml} \, T^{ij}{}_{kl},$$
$$T_{rs}{}^{tp} = g_{ri} \, g_{sj} \, g^{tk} g^{pl} \, T^{ij}{}_{kl}.$$

Note that the indices maintain their position relative to others.

When contracting over an index pair of a tensor product, you can simultaneously raise and lower the index pair:

$$A_j B^j = \left(g_{jk} A^k \right) \left(g^{jl} B_l \right) = g_{jk} \, g^{jl} A^k B_l = \delta^l_k A^k B_l = A^k B_k,$$
$$A_j B^j{}_r = \left(g_{jk} A^k \right) \left(g^{jl} B_{lr} \right) = g_{jk} \, g^{jl} A^k B_{lr} = \delta^l_k A^k B_{lr} = A^k B_{kr},$$
$$A^i{}_j B^j{}_r = \left(g_{jk} A^{ik} \right) \left(g^{jl} B_{lr} \right) = g_{jk} \, g^{jl} A^{ik} B_{lr} = \delta^l_k A^{ik} B_{lr} = A^{ik} B_{kr}.$$

A tensor is symmetric if it does not change by exchanging two indices. And thus the tensor equation $T_{kji} = T_{jki}$ express a (1,2) symmetry, symmetrical in the first and second index. A tensor is called *completely* symmetric if all permutations of the indices leave the tensor unchanged. For example the following (0,3)-tensor is completely symmetric:

$$T_{ijk} = T_{jki} = T_{kij} = T_{ikj} = T_{jik} = T_{kji}.$$

If a tensor is symmetrical in two indices, it loses its symmetry if only one of these indices is raised (lowered):

$$T_{kji} = T_{jki}, \quad (1,2) \text{ symmetry}$$
$$T_{kji} = g_{kl} T^l{}_{ji} = \underbrace{g_{kl} T^j{}_{li}}_{\textbf{no } \text{contraction over } l} \neq T^j{}_{ki}.$$

In the lower equation row, the (1,2) symmetry was applied in the second equation. Therefore, you must raise (lower) both symmetric indices in order for the tensor to maintain its symmetry:

$$T_{kji} = T_{jki}, \quad (1,2) \text{ symmetry}$$
$$T_{kji} = g_{kl}\, g_{jm} T^{lm}{}_i = \underbrace{g_{kl}\, g_{jm} T^{ml}{}_i}_{\text{contraction over } l, m} = T_{jki}.$$

A tensor is *antisymmetric* (or *skew-symmetric*) in two of its indices if the tensor changes its sign when these indices are exchanged. T_{kji} is antisymmetric in its first two indices, if $T_{kji} = -T_{jki}$.

A tensor can be symmetrized or anti-symmetrized over any number of its indices. The symmetrisation is done as follows: We add up all permutations of the chosen indices $(k_1 k_2 \cdots k_n)$, symbolised by round brackets, and divide by the number of terms:

$$\underbrace{T_{(k_1 k_2 \cdots k_n)}}_{\text{symbol}} = \underbrace{\frac{1}{n!} \left(T_{k_1 k_2 \cdots k_n} + \text{sum over all permutations of the indices } k_1 k_2 \cdots k_n \right)}_{\text{calculation}}.$$

$$(4.50)$$

With anti-symmetrisation over all indices, a completely anti-symmetric tensor, we form an alternating sum over the selected indices $[k_1 k_2 \cdots k_n]$, symbolised by square brackets, and assign a minus sign to the terms with odd permutation:

$$\underbrace{T_{[k_1 k_2 \cdots k_n]}}_{\text{symbol}} = \underbrace{\frac{1}{n!} \left(T_{k_1 k_2 \cdots k_n} + \text{alternating sum over all permutations of } k_1 k_2 \cdots k_n \right)}_{\text{calculation}}.$$

(4.51)

Example 4.5

An anti-symmetrisation of a (0,3)-tensor:

$$T_{[k_1 k_2 k_3]} = \frac{1}{3!} \left(T_{k_1 k_2 k_3} + T_{k_2 k_3 k_1} + T_{k_3 k_1 k_2} - T_{k_1 k_3 k_2} - T_{k_2 k_1 k_3} - T_{k_3 k_2 k_1} \right).$$

(4.52)

We want to show that if you make an index swap on places 1 and 3 equation (4.52) behaves anti-symmetrically:

$$\begin{aligned}
T_{[k_3 k_2 k_1]} &= \frac{1}{3!} \left(T_{k_3 k_2 k_1} + T_{k_1 k_3 k_2} + T_{k_2 k_1 k_3} - T_{k_2 k_3 k_1} - T_{k_3 k_1 k_2} - T_{k_1 k_2 k_3} \right) \\
&= -\frac{1}{3!} \left(-T_{k_3 k_2 k_1} - T_{k_1 k_3 k_2} - T_{k_2 k_1 k_3} + T_{k_2 k_3 k_1} + T_{k_3 k_1 k_2} + T_{k_1 k_2 k_3} \right) \\
&= -\frac{1}{3!} \left(T_{k_1 k_2 k_3} + T_{k_2 k_3 k_1} + T_{k_3 k_1 k_2} - T_{k_1 k_3 k_2} - T_{k_2 k_1 k_3} - T_{k_3 k_2 k_1} \right) \\
&= -T_{[k_1 k_2 k_3]}.
\end{aligned}$$

From the second to the third line, the terms were only rearranged.

∎

When two symmetric (antisymmetric) up-indices (or low) are contracted with low-indices (or up), only the symmetric (antisymmetric) part of the low-indices (or up) contributes:

$$S^{(kj)} T_{kj} = S^{(kj)} T_{(kj)}, \text{ or } A^{[kj]} T_{kj} = A^{[kj]} T_{[kj]}.$$

(4.53)

With a (1,1)-tensor T_j^k the contraction of upper index and lower index is called *trace* , which is a scalar. If you consider T_j^k as a matrix, the trace T is the sum of the diagonal entries:

$$T^i_i = \text{tr}(T).$$

(4.54)

With a (0,2)-tensor T_{kj} it is not allowed to contract the two lower indices together, because a contraction can only occur via an upper and lower index (see definition above). Therefore, an index must be raised, then you can contract afterwards:

$$\mathrm{tr}(T) = T^j_{\ j} = g^{jk} T_{kj}.$$

Let us take as an example the metric tensor g_{ji} for polar coordinates from equation (4.8), $g_{rr} = 1, g_{r\theta} = g_{\theta r} = 0, g_{\theta\theta} = r^2$. If we add the "diagonal components" we get $1 + r^2$. But this is not the correct value of the trace of the metric for polar coordinates! Raising or lowering it changes the numerical value of the components. And therefore the value of the trace $\mathrm{tr}(g_{ji})$ of our example is:

$$g^k_{\ i} = g^{kj}\, g_{ji},$$
$$\mathrm{tr}(g) = g^k_{\ k} = g^{kj}\, g_{jk}$$
$$= g^{rr}\, g_{rr} + g^{r\theta}\, g_{\theta r} + g^{\theta r}\, g_{r\theta} + g^{\theta\theta}\, g_{\theta\theta}$$
$$= 1 \cdot 1 + 0 + 0 + \frac{1}{r^2} \cdot r^2 = 2.$$

Box 4.2

A contraction is a tensor operation that has a number as its result, and is therefore frame invariant. We show this with the application of equation (3.17) and obtain:

$$T^{i'}_{\ i'} = \frac{\partial x^{i'}}{\partial x^i} \frac{\partial x^i}{\partial x^{i'}} T^i_{\ i} = R^{i'}_i F^i_{i'} T^i_{\ i} = T^i_{\ i}. \tag{4.55}$$

Chapter 5

Gradient

5.1 Differential operator

We introduce the differential operator $\tilde{\mathrm{d}}(\ )$:

$$\tilde{\mathrm{d}}(\ ) = \frac{\partial}{\partial x^i}(\ )\,\tilde{e}^i. \tag{5.1}$$

How does it differ from a differential d that we know from calculus? The differential dx is an infinitesimal length of the x-axis. In contrast, $\tilde{\mathrm{d}}(x)$ is a covector,

$$\tilde{\mathrm{d}}(x) = \frac{\partial x}{\partial x}\,\tilde{e}^x = 1 \cdot \tilde{e}^x. \tag{5.2}$$

In Chapter 2 we have already introduced covectors in detail. Figure 5.1 shows the pictorial difference between dx and $\tilde{\mathrm{d}}(x)$. And we remember the definition

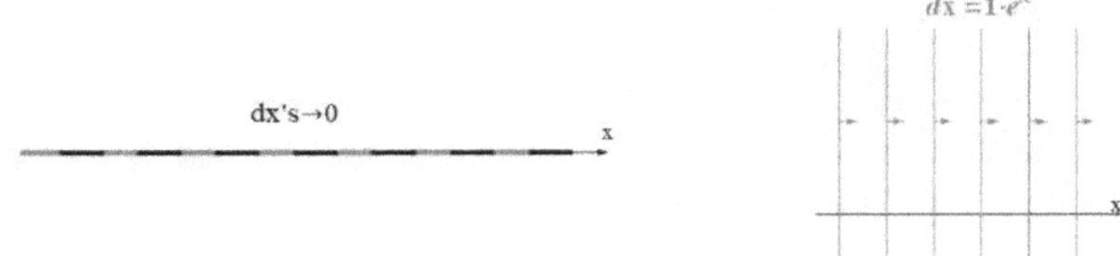

Figure 5.1: Pictorial difference between dx and $\tilde{\mathrm{d}}(x)$.

of a covector from Section 2.4: the lines are lines of constant value, and the direction of the covector always points to higher values.

5.2   Covector field or gradient

The differential d applied to a scalar field $f(x, y, z)$ yields by means of the chain rule:

$$df = \frac{\partial f}{\partial x}dx + \frac{\partial f}{\partial y}dy + \frac{\partial f}{\partial z}dz.$$

However, if we let the differential operator $\tilde{\mathrm{d}}(\ )$ act on this scalar field $f(x, y, z)$ so the result is a *covector field* or *one-form*:

$$\tilde{\mathrm{d}}(f) = \tilde{\mathrm{d}}f = \frac{\partial f}{\partial x}\tilde{e}^x + \frac{\partial f}{\partial y}\tilde{e}^y + \frac{\partial f}{\partial z}\tilde{e}^z = \frac{\partial f}{\partial x^i}\tilde{e}^i \tag{5.3}$$

For $\frac{\partial f}{\partial x^i}$ we write the short form, see equation (1.5), $\partial_i f$, and thus get

$$\tilde{\mathrm{d}}f = (\partial_i f)\tilde{e}^i, \tag{5.4}$$

which is a covector (one-form) and therefore maps a vector to a real number. His components are arranged as a row vector:

$$\tilde{\mathrm{d}}f = (\partial_i f)\tilde{e}^i \equiv \left(\frac{\partial f}{\partial x}, \frac{\partial f}{\partial y}, \frac{\partial f}{\partial z} \right). \tag{5.5}$$

This one-form or covector is called *gradient* of f.

Usually, in the three-dimensional vector calculus, the gradient is introduced as a vector. We'll talk about the difference between *covector gradient* and *vector gradient* later. The following Figure 5.2 shows a scalar field (the values increase from the blue area to the red area) and the associated gradient, the covector field. In the right picture of Figure 5.2 you can see the level set curves, the curves of constant value. The small arrows on the curves point in the direction of ascending values.

The components of a covector gradient $(\tilde{\mathrm{d}}f)_i$ transform just like the components of a one-form, see equation (2.12):

$$(\tilde{\mathrm{d}}f)_{j'} = (\partial_i f)\, F^i_{\ j'}, \quad F^i_{\ j'} = \frac{\partial x^i}{\partial x^{j'}}. \tag{5.6}$$

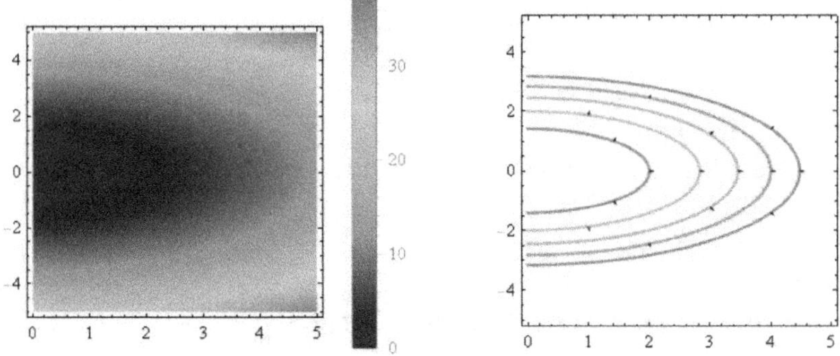

Figure 5.2: Scalar field and associated covector (one-form) gradient.

The *directional derivative*, the rate of change of a scalar field f along the direction (or directional velocity) $\vec{v}$ is defined by:

$$\tilde{\mathrm{d}}f(\vec{v}) = (\partial_i f)\tilde{e}^i(v^j \vec{e}_j) = (\partial_i f)v^j \tilde{e}^i(\vec{e}_j)$$
$$= (\partial_i f)v^j \delta^i_j = (\partial_i f)v^i. \tag{5.7}$$

If we let the one-form gradient $\tilde{\mathrm{d}}f(\ )$ act on a displacement vector $d\vec{s}$, we get the following known differential (directional derivative):

$$\tilde{\mathrm{d}}f(d\vec{s}) = (\partial_i f)\tilde{e}^i(dx^j \vec{e}_j) = (\partial_i f)dx^j \delta^i_j = (\partial_i f)dx^i = df.$$

The gradient $\tilde{\mathrm{d}}f$ of the scalar field f lives in dual space $\mathcal{V}^\star$. If we look for its equivalent from the vector space $\mathcal{V}$, the *vector gradient* $\vec{\mathrm{d}}f$, we obtain this with the help of the dual metric tensor $\boldsymbol{g}$, equation (4.43):

$$\vec{\mathrm{d}}f = \boldsymbol{g}\left(\tilde{\mathrm{d}}f;\ \right). \tag{5.8}$$

$\vec{\mathrm{d}}f$ is a *vector field* (contravariant components), unlike $\tilde{\mathrm{d}}f$, which is a covector field (covariant components). For $\vec{d}$ one often writes the 'del' symbol $\nabla \equiv \partial^i(\ )\vec{e}_i$:

$$\vec{d}f = \nabla f = \underbrace{\partial^i f}_{\text{contravariant}} \vec{e}_i. \tag{5.9}$$

And for the components we get:

$$\partial^j f = g^{ji}(\tilde{\mathrm{d}}f)_i,$$

$$\underbrace{\partial^j f}_{\text{contravariant}} = g^{ji} \underbrace{\partial_i f}_{\text{covariant}} . \tag{5.10}$$

The following simple example illustrates the difference between the components of covector gradient and vector gradient.

Example 5.1

For polar coordinates, the covector gradient components are:

$$(\partial_r f, \partial_\theta f) .$$

The polar metric tensor has the components, Section 4.3:

$$g_{rr} = 1, \ g_{r\theta} = g_{\theta r} = 0, \ g_{\theta\theta} = r^2.$$

And so we get for the dual metric tensor:

$$g^{rr} = 1, \ g^{r\theta} = g^{\theta r} = 0, \ g^{\theta\theta} = \frac{1}{r^2}.$$

The components of the vector gradient are thus:

$$\nabla f = \left(\partial^r f, \frac{1}{r^2} \partial^\theta f \right) . \tag{5.11}$$

And the vectorial representation (with coordinate basis!) is:

$$\nabla f = \partial^r f \, \vec{e}_r + \frac{1}{r^2} \partial^\theta f \, \vec{e}_\theta.$$

If we introduce a normalised basis $\{\vec{e}_r = \hat{e}_r, \vec{e}_\theta = r\hat{e}_\theta\}$, we recognise the familiar formula for a vector gradient for polar coordinates from vector calculus

$$\nabla f = \partial^r f \, \vec{e}_r + \frac{1}{r^2} \partial^\theta f \, \vec{e}_\theta = \partial^r f \, \hat{e}_r + \frac{1}{r^2} \partial^\theta f \, (r\hat{e}_\theta) = \partial^{\hat{r}} f \, \hat{e}_r + \frac{1}{r} \partial^{\hat{\theta}} f \, \hat{e}_\theta,$$

with $\left(\partial^{\hat{r}} f, \partial^{\hat{\theta}} f \right)$ as contravariant components of the vector gradient with orthonormal basis.

In this way, the vector gradient, or its components, of a scalar field f can be determined very quickly by working with the partial derivatives of the scalar field $\partial_i f$ (covariant components) and the inverse of the metric tensor $\mathbf{g}$ to get the contravariant components $\partial^j f$ of the vector gradient.   ∎

Box 5.1

The figural representation of the vector gradient ∇f is a directional vector that is perpendicular to a constant level line and points in the direction of increasing values. In this definition, we got to know the gradient in vector calculus.

We want to prove this with the tools of the past chapters. We start with equation (5.8):

$$\nabla f = \boldsymbol{g}\left(\tilde{\mathrm{d}}f; \right).$$

Let $\vec{v}$ be a vector of $\mathcal{V}$. With Definition 4.1 and equation (4.41) we get:

$$\nabla f \cdot \vec{v} = \mathbf{g}\left(;\nabla f, \vec{v}\right) = \tilde{\mathrm{d}}f(\vec{v}).$$

$\tilde{\mathrm{d}}f(\vec{v})$ is according to equation (5.7) the rate of change of f along $\vec{v}$. Now suppose that $\vec{t}$ is a tangent vector along a contour line. The rate of change along a contour is by definition zero. And thus:

$$\nabla f \cdot \vec{t} = 0.$$

Therefore the vector gradient ∇f is orthogonal to a contour line of f.

Chapter 6

Getting ready for curved space

General relativity (GR) uses a curved space-time to describe the curved trajectories of particles as an effect of gravity. To study the behaviour of vectors, basis vectors, one-forms etc. along curved paths, we first stay in flat space, where we are more confident in our approach. The mathematical results can almost be applied to a curved space (Section 6.6, 'local flatness theorem', Section 7.1, 'manifold').

A flat space is characterized by the fact that parallel lines never intersect (parallel axiom). In contrast, if the space is curved between two points, the idea that vectors point in the same direction at two different points may not have a definite solution. From now on we use the term *'manifold'* instead of 'space'. A manifold is a general topological space which behaves locally like a Euclidean space. More about this in the following sections.

6.1 Covariant derivative of a vector field

First we examine the ordinary derivative ∂_j according to a coordinate direction j of a $(1,0)$-tensor, e.g. the component of a vector, on its transformation behaviour. T^k be a (1,0)-tensor. We're executing a coordinate transformation:

$$\frac{\partial}{\partial x^{j'}} T^{k'} = \frac{\partial x^j}{\partial x^{j'}} \frac{\partial}{\partial x^j} \left(\frac{\partial x^{k'}}{\partial x^k} T^k \right)$$

$$= \frac{\partial x^j}{\partial x^{j'}} \frac{\partial x^{k'}}{\partial x^k} \left(\frac{\partial}{\partial x^j} T^k \right) + T^k \frac{\partial x^j}{\partial x^{j'}} \frac{\partial^2 x^{k'}}{\partial x^j \partial x^k}.$$

$$(6.1)$$

The first term on RHS of the second line shows the transformation behaviour of a (1,1)-tensor. The second term destroys this behaviour. Thus it is shown that the ordinary derivative does not in general transform a tensor back into a tensor, thereby violating the *covariance* between a physical relation and its transform. A derivation that preserves the covariance between the original object and the transformed object is called a *covariant derivation*. Thus a covariant derivation ensures the *independence* of physical equations from the chosen coordinates.

To distinguish the covariant derivative of a vector field $\vec{v}$ from the ordinary derivative $\partial_j v^i$, we introduce the "del" $\nabla_{()}(\ )$ symbol ($\nabla_{()}(\ ) \equiv$ **connection**; 'connection' because it defines and describes the progression through manifold M, Section 7.1, from point to point, from tangent space $\mathcal{T_P}$ to tangent space $\mathcal{T_Q}$, Section 7.2, with vector field as argument and a subindex that indicates the direction of the derivative:

$$\nabla_{\text{(directional derivative)}}(\text{input field}).$$

Box 6.1

The covariant derivative $\nabla_{()}(\ )$ is a linear mapping with the following properties:

1. It is linear in the first argument:

$$\nabla_{\partial_i + \partial_j} \vec{v} = \nabla_{\partial_i} \vec{v} + \nabla_{\partial_j} \vec{v},$$

$$\nabla_{a\,\partial_i} \vec{v} = a\nabla_{\partial_i} \vec{v}, \ a \in \mathbb{R}.$$

2. It is linear in the second argument:

$$\nabla_{\partial_i} (a\,\vec{v} + b\,\vec{w}) = a\,\nabla_{\partial_i} \vec{v} + b\,\nabla_{\partial_i} \vec{w}, \ a, b \in \mathbb{R}.$$

---

3. It meets the Leibniz rule:

$$\nabla_{\partial_i}\left(f\vec{v}\right) = \left(\partial_i f\right)\vec{v} + f\nabla_{\partial_i}\vec{v}, \ f \text{ is a smooth function.}$$

A covariant derivative of a vector field $\vec{v}$ in the direction of the k-coordinate (coordinate basis vector $\vec{e}_k$) or in the direction of vector field $\vec{u}$ is thus written as follows:

$$\nabla_{\vec{e}_k}\vec{v} = \nabla_{\partial_k}\vec{v}, \ \ \nabla_{\vec{u}}\vec{v}.$$

The covariant derivative describes the rate of change of a vector field or tensor field, taking into account that the basis vectors are *location-dependent* in magnitude and direction:

$$\nabla_{\partial_j}\left(\partial_i\right) \neq \vec{0}.$$

For a vector field with Cartesian coordinates (location-independent in magnitude and direction) or for a scalar field, the covariant derivative is identical to the ordinary partial derivative.

Figure 6.1 shows the transport of a vector field $\vec{v}$ along a vector field $\vec{u}$. A

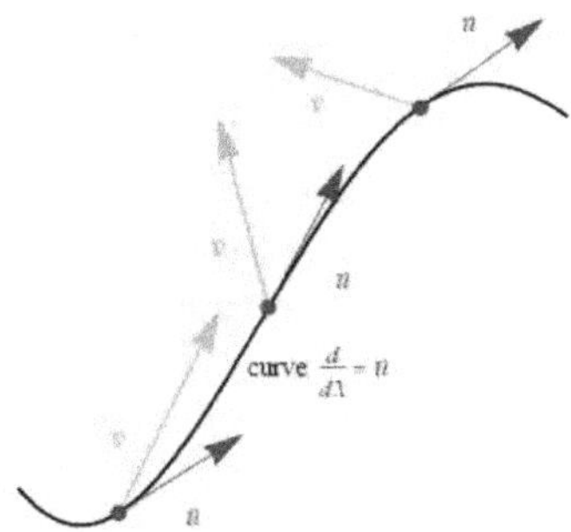

Figure 6.1: Transport of a vector field $\vec{v}$ along a vector field $\vec{u}$.

derivative is always a "movement" of an object (argument) in the direction of the derivative and after an infinitesimal movement the evaluation of the change of the object. The covariant derivative is defined as the differential limit of the transported vector $\vec{v}(\lambda)$ along a parametrized curve $\mathcal{C}(\lambda)$, with tangent vector

$\vec{u} = \frac{d}{d\lambda}$, after an infinitesimal step $d\lambda$ (for more details on the parametrization of a curve see Section 7.2):

$$\nabla_{\vec{u}} \vec{v} \equiv \lim_{d\lambda \to 0} \frac{\vec{v}(\lambda + d\lambda) - \vec{v}(\lambda)}{d\lambda}. \tag{6.2}$$

Remark 6.1
As we will learn later (Box 7.1, 'tangent space') only vectors of a manifold at the same point $\mathcal{P}$ can be compared and processed. So the vector $\vec{v}(\lambda + d\lambda)$ must be parallel transported back to point $\mathcal{P}(\lambda)$ to form a vector difference with $\vec{v}(\lambda)$. Equation (6.2) is only valid in a flat space, and serves here only for an easier understanding of the concept of covariant derivation.

■

In order to derive a calculative expression of the covariant derivative $\nabla_{\partial_i} \vec{v}$, we use the fact that "in the curved spacetime of general relativity the laws of physics have exactly the same mathematical form as they have in curved coordinate system in the flat spacetime of Minkowski space,"[1]. This fact justifies the following calculatory derivation for the expression of connection $\nabla_{()}()$. If we apply the differential rule separately for the components v^i and the basis vectors $\vec{e}_i$ for a vector field $\vec{v}$, we obtain the following calculus expression for the covariant derivative:

$$\nabla_{\partial_j} \vec{v} = \nabla_{\partial_j} (\vec{e}_i \, v^i) = \vec{e}_i \underbrace{\left(\frac{\partial}{\partial x^j} v^i \right)}_{\text{ordinary derivation}} + \underbrace{\left(\frac{\partial}{\partial x^j} \vec{e}_i \right) v^i}_{}. \tag{6.3}$$

$$\underbrace{\phantom{\nabla_{\partial_j} \vec{v} = \nabla_{\partial_j} (\vec{e}_i \, v^i) = \vec{e}_i \left(\frac{\partial}{\partial x^j} v^i \right) + \left(\frac{\partial}{\partial x^j} \vec{e}_i \right) v^i}}_{\text{vector field for a fixed } j}$$

The second term of the RHS of equation (6.3) describes the location dependence of the basis vectors. Since $\frac{\partial}{\partial x^j} \vec{e}_i$ itself is a vector per index pair, it can be written

---

[1][SS80, p218]

as a linear combination of the basis vectors:

$$\frac{\partial}{\partial x^j}\, \vec{e}_i = \Gamma^k_{\ ji}\, \vec{e}_k. \tag{6.4}$$

If we apply the Leibniz rule from Box 6.1 to $\nabla_{\partial_j}\vec{v}$, we get with the definition of equation (6.4)

$$\nabla_{\partial_j}\vec{e}_i = \Gamma^k_{\ ji}\, \vec{e}_k. \tag{6.5}$$

The symbol $\Gamma^k_{\ ji}$ gets its own name, the *Christoffel* symbol, or the *affine connection coefficient*. It should be noted that in this textbook the subscript indices of the Christoffel symbol are arranged in such a way that the first index defines the derivative direction and the second index denotes the basis vector to be processed.

Remark 6.2

1. The connection coefficient $\Gamma^k_{\ ji}$ as additive term in equation (6.3) provides the tensor properties of the covariant derivative. $\Gamma^k_{\ ji}$ itself is not a tensor and therefore it is not permitted to raise or lower the indices. Its transformation behaviour is given in Box 6.2.

2. In some texts the meaning of the subscript indices in the connection coefficients $\Gamma^k_{\ ji}$ is reversed: The first subscript index denotes the basis vector, and the second the direction of derivation.

3. The connection coefficients $\Gamma^k_{\ ji}$ are zero for Cartesian coordinates. The basic vectors $\{\vec{e}_x, \vec{e}_y, \ldots\}$ of Cartesian coordinates are location-independent in magnitude and direction.

∎

We can now write equation (6.3) more compactly: In the first term we relabel the dummy indices as follows, $i \leftrightarrow k$, and obtain:

$$\begin{aligned}
\nabla_{\partial_j}\vec{v} &= \vec{e}_k\left(\frac{\partial}{\partial x^j}\,v^k\right) + \vec{e}_k\,\Gamma^k_{\ ji}v^i \\[2mm]
&= \vec{e}_k\underbrace{\left(\frac{\partial}{\partial x^j}\,v^k + \Gamma^k_{\ ji}v^i\right)}_{k\text{-component for a fixed }j}
\end{aligned}$$

$$\nabla_{\partial_j}\vec{v} = \vec{e}_k\left(\partial_j\,v^k + \Gamma^k_{\ ji}v^i\right). \tag{6.6}$$

$\nabla_{\partial_j} \vec{v}$ describes a *tensor* field, and for a fixed j a *vector* field. The component form of the tensor field $\nabla_{\partial_j} \vec{v}$ is

$$\nabla_{\partial_j} v^k = \partial_j v^k + \Gamma^k_{ji} v^i. \tag{6.7}$$

And for a covariant derivative along a vector field $\vec{u}$ we get:

$$\nabla_{\vec{u}} \vec{v} = \nabla_{u^j \partial_j} \vec{v} \;=\; u^j \nabla_{\partial_j} \vec{v},$$

$$k\text{-component:}\; u^j \nabla_{\partial_j} v^k \;=\; u^j \left(\partial_j v^k + \Gamma^k_{ji} v^i \right). \tag{6.8}$$

Box 6.2

Transformation behaviour of the affine connection coefficient Γ^k_{ji}:

We start with equation (6.7)

$$\nabla_{\partial_j} v^k = \partial_j v^k + \Gamma^k_{ji} v^i. \tag{1}$$

$\nabla_{\partial_j} v^k$ is a tensor, therefore applies:

$$\nabla_{\partial_{j'}} v^{k'} = \frac{\partial x^j}{\partial x^{j'}} \frac{\partial x^{k'}}{\partial x^k} \nabla_{\partial_j} v^k. \tag{2}$$

We are transforming (1):

$$\nabla_{\partial_{j'}} v^{k'} = \partial_{j'} v^{k'} + \Gamma^{k'}_{j'i'} v^{i'} = \frac{\partial x^j}{\partial x^{j'}} \frac{\partial}{\partial x^j} \left(\frac{\partial x^{k'}}{\partial x^k} v^k \right) + \Gamma^{k'}_{j'i'} v^{i'}$$

$$= \frac{\partial x^j}{\partial x^{j'}} \frac{\partial x^{k'}}{\partial x^k} \left(\frac{\partial}{\partial x^j} v^k \right) + v^k \frac{\partial x^j}{\partial x^{j'}} \frac{\partial^2 x^{k'}}{\partial x^J \partial x^k} + \Gamma^{k'}_{j'i'} v^{i'}. \tag{3}$$

We include equation (1) in (2):

$$\nabla_{\partial_{j'}} v^{k'} = \frac{\partial x^j}{\partial x^{j'}} \frac{\partial x^{k'}}{\partial x^k} \nabla_{\partial_j} v^k = \frac{\partial x^j}{\partial x^{j'}} \frac{\partial x^{k'}}{\partial x^k} \left(\partial_j v^k + \Gamma^k_{ji} v^i \right)$$

$$= \frac{\partial x^j}{\partial x^{j'}} \frac{\partial x^{k'}}{\partial x^k} \partial_j v^k + \frac{\partial x^j}{\partial x^{j'}} \frac{\partial x^{k'}}{\partial x^k} \Gamma^k_{ji} v^i. \tag{4}$$

Because (3)=(4) and the first term of each equation is equal, we get:

$$\underbrace{v^k \frac{\partial x^j}{\partial x^{j'}}\frac{\partial^2 x^{k'}}{\partial x^J \partial x^k}}_{\text{dummy } i \leftrightarrow k} + \Gamma^{k'}_{j'i'} v^{i'} = \frac{\partial x^j}{\partial x^{j'}}\frac{\partial x^{k'}}{\partial x^k}\Gamma^k_{ji} v^i$$

$$\Gamma^{k'}_{j'i'}\frac{\partial x^{i'}}{\partial x^i} v^i = \frac{\partial x^j}{\partial x^{j'}}\frac{\partial x^{k'}}{\partial x^k}\Gamma^k_{ji} v^i - v^i \frac{\partial x^j}{\partial x^{j'}}\frac{\partial^2 x^{k'}}{\partial x^j \partial x^i}. \quad (5)$$

Equation (5) must be valid for each vector v^i. And therefore

$$\Gamma^{k'}_{j'i'} = \frac{\partial x^i}{\partial x^{i'}}\frac{\partial x^j}{\partial x^{j'}}\frac{\partial x^{k'}}{\partial x^k}\Gamma^k_{ji} - \frac{\partial x^i}{\partial x^{i'}}\frac{\partial x^j}{\partial x^{j'}}\frac{\partial^2 x^{k'}}{\partial x^j \partial x^i}.$$

The last term of equation (5) is identical with the additional term of the partial derivative, equation (6.1), with opposite sign, so that in combination ∂_j and Γ^k_{ji} show tensorial behaviour.

Remark 6.3

1. The covariant derivative $\nabla_{\partial_j}(\ )$ converts tensors to tensors. Thereby increasing the subindex by 1.

2. The covariant derivative $\nabla_{\partial_j}\vec{v}$ of vector field $\vec{v}$ is a vector field for a fixed j. $\nabla_{\partial_j} v^k$ are the components of a (1,1)-tensor.

■

Box 6.3

We want to determine the connection coefficients for polar coordinates. In Section 1.4 we have already calculated the polar basis vectors, and with the equations (1.9) and (1.10) the Christoffel values are given:

$$\frac{\partial \vec{e}_r}{\partial r} = \frac{\partial}{\partial r}\left(\cos\theta\,\vec{e}_x + \sin\theta\,\vec{e}_y\right) = 0 \implies \Gamma^j{}_{rr} = 0 \text{ for all } j,$$

$$\frac{\partial \vec{e}_r}{\partial \theta} = \frac{\partial}{\partial \theta}\left(\cos\theta\,\vec{e}_x + \sin\theta\,\vec{e}_y\right)$$

$$= -\sin\theta\,\vec{e}_x + \cos\theta\,\vec{e}_y = \frac{1}{r}\,\vec{e}_\theta \implies \Gamma^r{}_{\theta r} = 0,\ \Gamma^\theta{}_{\theta r} = \frac{1}{r},$$

$$\frac{\partial \vec{e}_\theta}{\partial r} = \frac{\partial}{\partial r}\left(-r\sin\theta\,\vec{e}_x + r\cos\theta\,\vec{e}_y\right)$$

$$= -\sin\theta\,\vec{e}_x + \cos\theta\,\vec{e}_y = \frac{1}{r}\,\vec{e}_\theta \implies \Gamma^r{}_{r\theta} = 0,\ \Gamma^\theta{}_{r\theta} = \frac{1}{r},$$

$$\frac{\partial \vec{e}_\theta}{\partial \theta} = \frac{\partial}{\partial \theta}\left(-r\sin\theta\,\vec{e}_x + r\cos\theta\,\vec{e}_y\right)$$

$$= -r\cos\theta\,\vec{e}_x - r\sin\theta\,\vec{e}_y = -r\,\vec{e}_r \implies \Gamma^r{}_{\theta\theta} = -r,\ \Gamma^\theta{}_{\theta\theta} = 0.$$

$$(6.9)$$

Using the equation (6.7) the covariant derivative for a vector field $\vec{v} = \left(v^r, v^\theta\right)$ in polar coordinates is obtained:

$$\nabla_{\partial_r}\vec{v} = \vec{e}_r\,\nabla_{\partial_r}v^r + \vec{e}_\theta\,\nabla_{\partial_r}v^\theta$$

$$= \vec{e}_r\left(\partial_r v^r + \Gamma^r{}_{rr}v^r + \Gamma^r{}_{r\theta}v^\theta\right) + \vec{e}_\theta\left(\partial_r v^\theta + \Gamma^\theta{}_{rr}v^r + \Gamma^\theta{}_{r\theta}v^\theta\right)$$

$$= \vec{e}_r\,\partial_r v^r + \vec{e}_\theta\left(\partial_r v^\theta + \frac{1}{r}v^\theta\right).$$

$$\nabla_{\partial_\theta}\vec{v} = \vec{e}_r\,\nabla_{\partial_\theta}v^r + \vec{e}_\theta\,\nabla_{\partial_\theta}v^\theta$$

$$= \vec{e}_r\left(\partial_\theta v^r + \Gamma^r{}_{\theta r}v^r + \Gamma^r{}_{\theta\theta}v^\theta\right) + \vec{e}_\theta\left(\partial_\theta v^\theta + \Gamma^\theta{}_{\theta r}v^r + \Gamma^\theta{}_{\theta\theta}v^\theta\right)$$

$$= \vec{e}_r\left(\partial_\theta v^r - rv^\theta\right) + \vec{e}_\theta\left(\partial_\theta v^\theta + \frac{1}{r}v^r\right).$$

$$(6.10)$$

Determining the connection coefficients $\Gamma^k{}_{ji}$ in the way shown here in Box 6.3 can be very laborious for other coordinates. We have done it here for better understanding. A simpler method for determining the connection coefficients is shown in Section 6.4 using the formula of the Levi-Civita connection, equation (6.25).

6.2 Covariant derivative and coordinate Transformation

In order to determine the components $\nabla_{\partial_j} v^k$ of a covariant derivative after a coordinate transformation $\nabla_{\partial_{j'}} v^{k'}$, there are two ways. First, using equation (6.7) and known Γ^k_{ji}, Example 6.1, or in the more elegant way by applying the tensor transformation rules, equation (3.17), see Example 6.2.

Example 6.1

The covariant derivatives with respect to the polar coordinates $\{r, \theta\}$ of vector field $\vec{v}$ in Cartesian coordinates $\vec{v} = x\,\vec{e}_x + y\,\vec{e}_y$ are to be determined. We first calculate the polar components v^r and v^θ.

$$\vec{v} = x\,\vec{e}_x + y\,\vec{e}_y,$$
$$x = r\,\cos\theta, \quad y = r\,\sin\theta,$$
$$\vec{e}_x = \cos\theta\,\vec{e}_r - \frac{\sin\theta}{r}\,\vec{e}_\theta,$$
$$\vec{e}_y = \sin\theta\,\vec{e}_r + \frac{\cos\theta}{r}\,\vec{e}_\theta.$$

And so we get:

$$\vec{v}_{\text{polar}} = r\,\cos\theta\left(\cos\theta\,\vec{e}_r - \frac{\sin\theta}{r}\,\vec{e}_\theta\right) + r\,\sin\theta\left(\sin\theta\,\vec{e}_r + \frac{\cos\theta}{r}\,\vec{e}_\theta\right)$$
$$= \left(r\,\cos^2\theta + r\,\sin^2\theta\right)\vec{e}_r + \left(-\cos\theta\,\sin\theta + \sin\theta\,\cos\theta\right)\vec{e}_\theta$$
$$= r\,\vec{e}_r.$$

The result clearly shows that the vector field $\vec{v}$ is a radial one as we expected. The covariant derivative of the vector field $\vec{v}$ in polar coordinates is obtained with the Christoffel symbols from equation (6.10)

$$\nabla_{\partial_r}\vec{v} = \vec{e}_r, \quad \nabla_{\partial_\theta}\vec{v} = \vec{e}_\theta$$

■

Example 6.2

We want to approach the transition of a covariant derivative from the Cartesian coordinate system $\nabla_{\partial_j} v^i$ to the polar coordinate system $\nabla_{\partial_{j'}} v^{i'}$ with the help of the tensor transformation rules equation (3.17). Our vector field is still that from Example 6.1: $\vec{v} = x\,\vec{e}_x + y\,\vec{e}_y$. We apply equation (3.17):

$$\nabla_{\partial_{j'}} v^{i'} = R_i^{i'} \left(\nabla_{\partial_j} v^i\right) F_{j'}^{j}.$$

The unprimed indices refer to Cartesian coordinates, and therefore:

$$\nabla_{\partial_j} v^i = \partial_j v^i = \begin{pmatrix} \frac{\partial v^x}{\partial x} & \frac{\partial v^x}{\partial y} \\ \frac{\partial v^y}{\partial x} & \frac{\partial v^y}{\partial y} \end{pmatrix} = \begin{pmatrix} 1 & 0 \\ 0 & 1 \end{pmatrix}.$$

With $R_i^{i'}$ and $F_{j'}^{j}$ from equation (1.18) we get:

$$\begin{aligned}
\nabla_{\partial_{j'}} v^{i'} &= R_i^{i'} \left(\nabla_{\partial_j} v^i\right) F_{j'}^{j} \\
&= \begin{pmatrix} \cos\theta & \sin\theta \\ \frac{-\sin\theta}{r} & \frac{\cos\theta}{r} \end{pmatrix} \begin{pmatrix} 1 & 0 \\ 0 & 1 \end{pmatrix} \begin{pmatrix} \cos\theta & -r\sin\theta \\ \sin\theta & r\cos\theta \end{pmatrix} = \begin{pmatrix} 1 & 0 \\ 0 & 1 \end{pmatrix} \\
&= \begin{pmatrix} \nabla_{\partial_r} v^r & \nabla_{\partial_\theta} v^r \\ \nabla_{\partial_r} v^\theta & \nabla_{\partial_\theta} v^\theta \end{pmatrix}.
\end{aligned}$$

And therefore we obtain

$$\nabla_{\partial_r} \vec{v} = \vec{e}_r\,\nabla_{\partial_r} v^r + \vec{e}_\theta\,\nabla_{\partial_r} v^\theta = \vec{e}_r \cdot 1 + \vec{e}_\theta \cdot 0 = \vec{e}_r,$$
$$\nabla_{\partial_\theta} \vec{v} = \vec{e}_r\,\nabla_{\partial_\theta} v^r + \vec{e}_\theta\,\nabla_{\partial_\theta} v^\theta = \vec{e}_r \cdot 0 + \vec{e}_\theta \cdot 1 = \vec{e}_\theta.$$

$\blacksquare$

6.3   Covariant derivative of a covector field

We may well assume that a covariant derivative of a covector displays a resemblance to that of a vector. From this basic consideration we generate a scalar field f, which has been created by means of contraction from the processing of a vector $\vec{v}$ by a covector $\tilde{p}$: $f = \tilde{p}(\vec{v})$. If we then apply an ordinary derivative to the scalar field f and then compare it with the covariant derivative of f, we can derive an expression for a covariant derivative of a covector as shown below.

$$\tilde{p}(\vec{v}) = p_j \tilde{e}^j \left(v^k \vec{e}_k\right) = p_j v^k \tilde{e}^j (\vec{e}_k) = p_j v^k \delta_j^k = p_j v^j = f. \tag{6.11}$$

We differentiate equation 6.11 according to the product rule of scalar values and a second a covariant derivative of the covector / vector components p_j and v^j. First:

$$\partial_i f = v^j \, \partial_i p_j + p_j \, \partial_i v^j. \tag{6.12}$$

And second:

$$\nabla_{\partial_i} f = \partial_i f = v^j \, \nabla_{\partial_i} p_j + p_j \, \nabla_{\partial_i} v^j. \tag{6.13}$$

According to equation (6.7) $\partial_i v^j$ is:

$$\partial_i v^j = \nabla_{\partial_i} v^j - \Gamma^j_{ik} v^k. \tag{6.14}$$

With this result we enter equation (6.12):

$$\begin{aligned}
\partial_i f &= v^j \partial_i p_j + p_j \left(\nabla_{\partial_i} v^j - \Gamma^j_{ik} v^k \right) \\
&= v^j \partial_i p_j + p_j \nabla_{\partial_i} v^j - \underbrace{p_j \, \Gamma^j_{ik} v^k}_{j \longrightarrow k, k \longrightarrow j} \\
&= v^j \partial_i p_j - p_k \, \Gamma^k_{ij} v^j + p_j \, \nabla_{\partial_i} v^j \\
&= v^j \left(\partial_i p_j - p_k \, \Gamma^k_{ij} \right) + p_j \, \nabla_{\partial_i} v^j.
\end{aligned} \tag{6.15}$$

We compare this expression with equation (6.13) and thereby obtain the result for the covariant derivative of a covector:

$$\nabla_{\partial_i} \tilde{p} = \left(\partial_i p_j - p_k \, \Gamma^k_{ij} \right) \tilde{e}^j,$$
$$j\text{-component: } \nabla_{\partial_i} p_j = \partial_i p_j - p_k \, \Gamma^k_{ij}. \tag{6.16}$$

If you compare the two formulas for the covariant derivative of a vector equation (6.7) and of a covector equation (6.16)

$$\nabla_{\partial_i} v^j = \partial_i v^j + \Gamma^j_{ik} v^k,$$
$$\nabla_{\partial_i} p_j = \partial_i p_j - p_k \, \Gamma^k_{ij}, \tag{6.17}$$

you can see a change of sign in front of the Christoffel symbol. The reason for this is that basis vectors and basis covectors show opposite behaviour

Remark 6.4

The covariant derivative of a vector $\nabla_{\partial_i} v^j$ or a one-form $\nabla_{\partial_i} p_j$ is a tensor:

$$\nabla_{\partial_{i'}} v^{j'} = R^{j'}_m F^n_{i'} \nabla_{\partial_n} v^m = R^{j'}_m F^n_{i'} \left(\partial_n v^m + \Gamma^m_{nk} v^k \right),$$
$$\nabla_{\partial_{i'}} p_{j'} = F^m_{j'} F^n_{i'} \nabla_{\partial_n} p_m = F^m_{j'} F^n_{i'} \left(\partial_n p_m - p_k \, \Gamma^k_{nm} \right).$$

The ordinary derivation of a one-form $\partial_i p_j$, however, does not result in a tensor. The proof proceeds in a similar way to that for a $(1,0)$-tensor as shown in Section 6.1.

■

Box 6.4

The formulas in equation (6.17) can be used to derive covariant derivatives of higher ranking tensors. The following applies to rank 2 tensors:

$$(2,0)\text{-tensor: } \nabla_{\partial_i} A^{jk} = \partial_i A^{jk} + \Gamma^{j}_{\ il} A^{lk} + \Gamma^{k}_{\ il} A^{jl},$$

$$(1,1)\text{-tensor: } \nabla_{\partial_i} B^{j}_{\ k} = \partial_i B^{j}_{\ k} + \Gamma^{j}_{\ il} B^{l}_{\ k} - B^{j}_{\ l}\Gamma^{l}_{\ ik}, \qquad (6.18)$$

$$(0,2)\text{-tensor: } \nabla_{\partial_i} C_{jk} = \partial_i C_{jk} - C_{lk}\Gamma^{l}_{\ ij} - C_{jl}\Gamma^{l}_{\ ik}.$$

A small rule of thumb: the starting indices (LHS) must be present in every term on the RHS after contraction.

In the course of the textbook we will encounter the $(0,2)$ metric tensor g_{ij} again and again. Therefore, we want to record its covariant derivative at this point.

$$\nabla_{\partial_i} g_{sj} = \partial_i g_{sj} - g_{lj}\Gamma^{l}_{\ is} - g_{sl}\Gamma^{l}_{\ ij}. \qquad (6.19)$$

6.4   Connection and metric

In Section 5.2 we have shown that the differential operator $\tilde{\mathrm{d}}$ applied to a scalar field f produces a co-vector field (one-form, gradient) with the components $\partial_i f$, equation (5.4). When applied a second time, we get a $(0, 2)$-tensor with the components $\nabla_{\partial_j} (\partial_i f)$. In Cartesian coordinates: $\nabla_{\partial_j} (\partial_i f) = \partial_j (\partial_i f)$. At this point we must again clarify that the partial derivative ∂_i has the same properties as a coordinate basis vector $\vec{e}_i$. Compare also the explanations in Section 1.4. In Section 7.10 we will show that ∂_i always commute $[\partial_i, \partial_j] = 0$, unlike a non-coordinate basis where this is not guaranteed. For this reason we may formulate the following:

$$\nabla_{\partial_j} (\partial_i f) = \partial_j (\partial_i f) = \partial_i (\partial_j f) = \nabla_{\partial_i} (\partial_j f).$$

If a tensor is symmetric in one coordinate system it is in all, Definition 3.1, and therefore:

$$\nabla_{\partial_j}\left(\partial_i f\right) = \nabla_{\partial_i}\left(\partial_j f\right),$$
$$\partial_j\left(\partial_i f\right) - \left(\partial_k f\right)\Gamma^k_{ji} = \partial_i\left(\partial_j f\right) - \left(\partial_k f\right)\Gamma^k_{ij}, \tag{6.20}$$
$$\left(\partial_k f\right)\Gamma^k_{ji} = \left(\partial_k f\right)\Gamma^k_{ij}.$$

Thus, the Christoffel symbol Γ, the connection coefficient, is symmetrical in its sub-indices:

$$\Gamma^k_{ji} = \Gamma^k_{ij}. \tag{6.21}$$

Equation (6.21) is only valid for a *coordinate basis*, but *not* for a non-coordinate basis, see also Sections 7.10, 7.12 and 7.13.

We now want to derive a relationship between the connection ∇, the connection coefficient Γ and the metric tensor $\mathbf{g}$. The terms used here (manifold, tangent space, etc.) will be discussed in Chapter 7. Nevertheless it should not be difficult to follow the explanations and derivations. First we need to define the concept of metric compatibility:

Definition 6.1   *metric compatibility*
A connection ∇ on a manifold M is called *compatible with the metric* $\mathbf{g}$ on M, if for any two vector fields $\vec{U}$ and $\vec{V}$ on M and any tangent vector $\vec{v}$ of the tangent space $\mathcal{T}_P$ is valid:

$$\vec{v}\left(\mathbf{g}\left(\vec{U},\vec{V}\right)\right) = \mathbf{g}\left(\nabla_{\vec{v}}\vec{U},\vec{V}\right) + \mathbf{g}\left(\vec{U},\nabla_{\vec{v}}\vec{V}\right). \tag{6.22}$$

On the LHS a smooth function $\mathbf{g}\left(\vec{U},\vec{V}\right)$ is derived by vector $\vec{v}$. This is a well-defined expression *without* reference to any connection. The RHS is dependent on the connection $\nabla_{\vec{v}}$.

We apply equation (6.22) to the coordinates basis vectors:

$$\partial_i\left(\vec{e}_s \cdot \vec{e}_j\right) = \left(\nabla_{\partial_i}\vec{e}_s\right)\cdot\vec{e}_j + \vec{e}_s\cdot\left(\nabla_{\partial_i}\vec{e}_j\right). \tag{6.23}$$

If a metric compatibility in the sense of the Definition 6.1 is given, we obtain the following expression with equation (6.23) for the metric tensor g_{sj}, Definition

4.1, equation (4.6):

$$\partial_i g_{sj} = \partial_i \left(\vec{e}_s \cdot \vec{e}_j\right) = \left(\nabla_{\partial_i} \vec{e}_s\right) \cdot \vec{e}_j + \vec{e}_s \cdot \left(\nabla_{\partial_i} \vec{e}_j\right)$$

$$= \left(\vec{e}_l \, \Gamma^l_{is}\right) \cdot \vec{e}_j + \vec{e}_s \cdot \left(\vec{e}_l \, \Gamma^l_{ij}\right)$$

$$\underset{\Gamma \text{ is a scalar}}{=} \vec{e}_l \cdot \vec{e}_j \, \Gamma^l_{is} + \vec{e}_s \cdot \vec{e}_l \, \Gamma^l_{ij} = \underbrace{g_{lj}}_{=g_{jl}} \Gamma^l_{is} + g_{sl} \, \Gamma^l_{ij}$$

$$(1) \quad \partial_i g_{sj} = g_{jl} \, \Gamma^l_{is} + g_{sl} \, \Gamma^l_{ij}. \tag{6.24}$$

We now shift the indices of the LHS to the right by one: $i\,s\,j \longrightarrow j\,i\,s$, and get:

$$(2) \quad \partial_i g_{sj} = g_{jl} \, \Gamma^l_{is} + g_{sl} \, \Gamma^l_{ij}.$$

Shifting the LHS indices a second time to the right, $j\,i\,s \longrightarrow s\,j\,i$,

$$(3) \quad \partial_s g_{ji} = g_{il} \, \Gamma^l_{sj} + g_{jl} \, \Gamma^l_{si}.$$

Due to the symmetry of the connections Γ, we recognize that the first term of (1) is identical to the second term of (3), likewise the second term of (1) with the first term of (2), and the second term of (2) with the first term of (3). With the following combination of the three equations, (1)+(2)-(3), we get:

$$\partial_i \, g_{sj} + \partial_j \, g_{is} - \partial_s \, g_{ji} = 2 g_{sl} \, \Gamma^l_{ji}$$

$$g^{sk} \left(\partial_i \, g_{sj} + \partial_j \, g_{is} - \partial_s \, g_{ji}\right) = 2 \underbrace{g_{sl} g^{sk}}_{=\delta^k_l} \, \Gamma^l_{ji}.$$

And thus we get the following important expression for the *affine connection*

$$\Gamma^k_{ji} = \frac{1}{2} g^{ks} \left(\partial_j \, g_{is} + \partial_i \, g_{sj} - \partial_s \, g_{ji}\right), \tag{6.25}$$

which represents the relationship between the Christoffel symbol and the partial derivatives of the metric tensor. This relationship is also called the *Levi-Civita connection* or the *Riemann connection*.

With the relations equation (6.19): $\nabla_{\partial_i} g_{sj} = \partial_i g_{sj} - g_{lj}\Gamma^l_{is} - g_{sl}\Gamma^l_{ij}$, and (6.24): $\partial_i g_{sj} = g_{jl} \, \Gamma^l_{is} + g_{sl} \, \Gamma^l_{ij}$, it results that the covariant derivative of the metric g_{ij} is identically zero:

$$\nabla_{\partial_i} g_{sj} = 0. \tag{6.26}$$

Therefore, *the metric is constant in the direction of each smooth vector field.*

Example 6.3

We already determined the Christoffel symbols for polar coordinates in Box 6.3. But let us now do this with the Levi-Civita connection, equation (6.25) we have just obtained. We take the values for g_{ij} and g^{kl} from Example 5.1.

$$
\begin{aligned}
\Gamma^\theta{}_{\theta r} &= \frac{1}{2}\, g^{\theta s}\left(\partial_\theta\, g_{rs} + \partial_r\, g_{s\theta} - \partial_s\, g_{\theta r}\right) \\
&= \frac{1}{2}\left\{ g^{\theta r}\left(\partial_\theta\, g_{rr} + \partial_r\, g_{r\theta} - \partial_r\, g_{\theta r}\right) + g^{\theta\theta}\left(\partial_\theta\, g_{r\theta} + \partial_r\, g_{\theta\theta} - \partial_\theta\, g_{\theta r}\right) \right\} \\
&= \frac{1}{2}\left\{ 0\cdot\left(\partial_\theta\, g_{rr} + \partial_r\, g_{r\theta} - \partial_r\, g_{\theta r}\right) + \frac{1}{r^2}\left(0 + 2r - 0\right) \right\} \\
&= \frac{1}{r}.
\end{aligned}
$$

$$
\begin{aligned}
\Gamma^r{}_{\theta\theta} &= \frac{1}{2}\, g^{rs}\left(\partial_\theta\, g_{\theta s} + \partial_\theta\, g_{s\theta} - \partial_s\, g_{\theta\theta}\right) \\
&= \frac{1}{2}\left\{ g^{rr}\left(\partial_\theta\, g_{\theta r} + \partial_\theta\, g_{r\theta} - \partial_r\, g_{\theta\theta}\right) + g^{r\theta}\left(\partial_\theta\, g_{\theta\theta} + \partial_\theta\, g_{\theta\theta} - \partial_\theta\, g_{\theta\theta}\right) \right\} \\
&= \frac{1}{2}\left\{ 1\cdot\left(0 + 0 - 2r\right) + 0\cdot\left(\partial_\theta\, g_{\theta\theta} + \partial_\theta\, g_{\theta\theta} - \partial_\theta\, g_{\theta\theta}\right) \right\} \\
&= -r.
\end{aligned}
$$

And so on.

■

6.5 Divergence and Laplacian

We want to introduce another abbreviated notation: $\nabla_{\partial_j} \equiv \nabla_j$.

The *divergence* of a vector field is a *scalar field* that indicates the source strength at each point. This scalar field is *frame-invariant*. Which is why we are concluding: In Cartesian coordinates, the divergence for a vector field $\vec{v} = \vec{e}_i v^i$ is equal to $\partial_i v^i$. $\partial_i v^i$ means the tensor operation 'contraction' . This is frame invariant. Therefore applies:

$$
\partial_i v^i = \nabla_j v^j = \operatorname{div}\vec{v}. \tag{6.27}
$$

A small example in advance before we derive a compact formula for the divergence with metric compatibility. The divergence of a vector field in polar coordinates results directly from equation (6.10):

$$\nabla_j v^j = \nabla_r v^r + \nabla_\theta v^\theta = \left(\frac{\partial v^r}{\partial r}\right) + \left(\frac{\partial v^\theta}{\partial \theta} + \frac{1}{r} v^r\right) = \frac{1}{r}\frac{\partial}{\partial r}\left(r\, v^r\right) + \frac{\partial v^\theta}{\partial \theta}. \quad (6.28)$$

Remark 6.5

Equation (6.28) looks somewhat different from what we know from vector calculus. This is because there one generally works with a normalised basis $\hat{e}_i$. Equation (6.28), however, represents the divergence in polar coordinates for a *coordinate basis*, which is generally used in this textbook. For an orthonormal basis, the expression of equation (6.28) changes to:

$$\nabla_{\hat{j}} v^{\hat{j}} = \frac{1}{r}\frac{\partial}{\partial r}\left(r\, v^{\hat{r}}\right) + \frac{1}{r}\frac{\partial v^{\hat{\theta}}}{\partial \theta}.$$

$\blacksquare$

In general, the divergence of a vector field v^k is defined by equation (6.7) (note the identical indices of contraction):

$$\nabla_k v^k = \partial_k v^k + \Gamma^k_{ki} v^i. \quad (6.29)$$

The Christoffel connection in the divergence form Γ^k_{ki} is with equation (6.25):

$$\begin{aligned}
\Gamma^k_{ki} &= \frac{1}{2} g^{ks}\left(\partial_k g_{is} + \partial_i g_{sk} - \partial_s g_{ki}\right) \\
&= \frac{1}{2} g^{ks}\left(\partial_k g_{is} - \partial_s g_{ki}\right) + \frac{1}{2} g^{ks}\partial_i g_{sk}. \quad (6.30)
\end{aligned}$$

The first term of the second line of the above equation yields zero, since the contraction via the indices k and s is a contraction via a symmetrical tensor, g^{ks}, and an anti-symmetrical term, $(\partial_k g_{is} - \partial_s g_{ki})$. And this adds up to zero, because the symmetric tensor, g^{ks}, selects all symmetric (k, s)-parts of $(\partial_k g_{is} - \partial_s g_{ki})$, which are not existent because of the asymmetry in these indices. See Section 4.11, equation (4.53). This reduces the connection to

$$\Gamma^k_{ki} = \frac{1}{2} g^{ks}\partial_i g_{sk}. \quad (6.31)$$

If we take a closer look at this expression, we notice that we have a derivative of a quantity g_{sk}, multiplied by its inverse g^{ks}. And this is the derivative of the natural logarithm. The conversion of (6.31) into a more compact form is shown in Box 6.5.

Box 6.5

Starting from the Laplace expansion to calculate the determinant of a matrix, Appendix, Section B.1, (B.6),

$$\det (A) = \sum_j a^i_j C^i_j, \text{ for any fixed } i,$$

we differentiate it with respect to the matrix element a^i_j:

$$\frac{\partial (\det (A))}{\partial a^i_j} = C^i_j \stackrel{(B.7)}{=} \det(A)(A^{-1})^j_i.$$

And with the use of the chain rule we get

$$\frac{1}{\det(A)} \frac{\partial \det(A)}{\partial x} = (A^{-1})^j_i \frac{\partial a^i_j}{\partial x},$$

$$\partial_x (\ln \det(A)) = (A^{-1})^j_i \partial_x a^i_j. \tag{6.32}$$

The RHS of (6.32) corresponds to the RHS of equation (6.31). We choose the short form $g_{sk} \equiv g$ and therefore receive for Γ^k_{ki} of (6.31):

$$\frac{1}{2}\partial_i \ln(\det g) = \partial_i \ln \sqrt{|\det g|} = \frac{1}{\sqrt{|\det g|}} \partial_i \sqrt{|\det g|},$$

$$\Rightarrow$$

$$\Gamma^k_{ki} = \frac{1}{\sqrt{|\det g|}} \partial_i \sqrt{|\det g|}.$$

And the divergence of equation (6.29) results in:

$$\nabla_k v^k = \partial_k v^k + \underbrace{\Gamma^k_{ki} v^i}_{k \longrightarrow j,\, i \longrightarrow k} = \partial_k v^k + \Gamma^j_{jk} v^k$$

$$= \partial_k v^k + \frac{1}{\sqrt{|\det g|}} \partial_k \sqrt{|\det g|}\, v^k = \frac{1}{\sqrt{|\det g|}} \partial_k \left(\sqrt{|\det g|} v^k \right)$$

$$\nabla_k v^k = \frac{1}{\sqrt{|\det g|}} \partial_k \left(\sqrt{|\det g|} v^k \right). \tag{6.33}$$

Remark 6.6

Note that equation (6.33) leads to different results than those known from the Vector Calculus textbooks, because this formula is based on coordinate basis and not on orthonormal basis vectors. ∎

Example 6.4

We want to determine the divergence of a vector field $\vec{v}$ in a spherical coordinate system. The matrix representation of the metric tensor g_{sk}, see Example 4.3, is

$$
(g_{sk}) = \begin{pmatrix} 1 & 0 & 0 \\ 0 & r^2 & 0 \\ 0 & 0 & r^2 \sin^2 \theta \end{pmatrix},
$$

and the determinant therefore

$$
\det(g_{sk}) = r^4 \sin^2 \theta, \;\Rightarrow\; \sqrt{|\det g|} = r^2 \sin \theta.
$$

And with equation (6.33), *coordinate basis*, we get:

$$
\mathrm{div}\vec{v} = \nabla_k v^k = \frac{1}{r^2 \sin \theta} \left[\partial_r \left(r^2 \sin \theta\, v^r \right) + \partial_\theta \left(r^2 \sin \theta\, v^\theta \right) + \partial_\phi \left(r^2 \sin \theta\, v^\phi \right) \right].
$$

In the case of an *orthonormal basis*, the formula is

$$
\mathrm{div}\vec{v} = \nabla_{\hat{k}} v^{\hat{k}} = \frac{1}{r^2 \sin \theta} \left[\partial_{\hat{r}} \left(r^2 \sin \theta\, v^{\hat{r}} \right) + \partial_{\hat{\theta}} \left(r \sin \theta\, v^{\hat{\theta}} \right) + \partial_{\hat{\phi}} \left(r\, v^{\hat{\phi}} \right) \right].
$$

Note the index designation of the components, which now refer to an orthonormal basis.

■

The *Laplacian* ∇^2 is defined as the divergence of a *vector gradient* of a scalar field f:

$$
\nabla^2 f = \nabla \cdot \nabla f = \nabla_j g^{ji} \partial_i f \tag{6.34}
$$

The Laplacian for polar coordinates is calculated in a straightforward manner as follows. The polar components $(\nabla f)^j = g^{ji} \partial_i f$ of a vector gradient have been determined in Example 5.1: $\nabla f = \left(\partial^r f, \frac{1}{r^2} \partial^\theta f \right)$. We enter these components into equation (6.28) and thus obtain the Laplacian of the scalar field f in polar coordinates:

$$
\nabla^2 f = \frac{1}{r} \frac{\partial}{\partial r} \left(r \frac{\partial f}{\partial r} \right) + \frac{1}{r^2} \frac{\partial^2 f}{\partial \theta^2}.
$$

6.6 Local flatness theorem

In Section 4.4 we have shown in principle that the metric can be put into a canonical form, with all the advantages of a Euclidean structure and its orthonormal basis (Cartesian coordinates). A coordinate basis is orthogonal, but generally not orthonormal. However, it can be normalized by rescaling (see Example 6.5).

Now we want to prove that a canonization of the metric is always possible at a point $\mathcal{P}$ of a vector space ($n = 4$ and $n = 3$). The following investigation of the position dependence of $\mathbf{g}$ during coordinate transformation for a vector space of $n = 4$ (or $n = 3$) using a Taylor expansion shows the following result.

$$
\begin{aligned}
g'(\vec{x}) &= g'\Big|_{\mathcal{P}} + \Delta x^{k'} \left(\partial_{k'} g'\right)\Big|_{\mathcal{P}} + \frac{1}{2}\Delta x^{l'}\Delta x^{k'} \left(\partial_{l'}\partial_{k'} g'\right)\Big|_{\mathcal{P}} + \cdots \quad (6.35) \\
&= \frac{\partial x^i}{\partial x^{i'}}\frac{\partial x^j}{\partial x^{j'}} g_{ij}\Big|_{\mathcal{P}} + \Delta x^{k'} \left(\frac{\partial}{\partial x^{k'}}\left(\frac{\partial x^i}{\partial x^{i'}}\frac{\partial x^j}{\partial x^{j'}} g_{ij}\right)\right)\Big|_{\mathcal{P}} + \quad (6.36) \\
&\quad + \frac{1}{2}\Delta x^{l'}\Delta x^{k'} \left(\frac{\partial}{\partial x^{l'}}\frac{\partial}{\partial x^{k'}}\left(\frac{\partial x^i}{\partial x^{i'}}\frac{\partial x^j}{\partial x^{j'}} g_{ij}\right)\right)\Big|_{\mathcal{P}} + \cdots
\end{aligned}
$$

We compare free components of the different terms of the equation (6.35) and (6.36).

First term.
(6.35): $g'\Big|_{\mathcal{P}}$ is an array of 10 numbers (symmetrical $(0, 2)$-tensor). For $n = 3 : 6$ numbers.
(6.36): The transformation matrix $F^i_{i'} = \frac{\partial x^i}{\partial x^{i'}}$ of $\frac{\partial x^i}{\partial x^{i'}}\frac{\partial x^j}{\partial x^{j'}} g_{ij}\Big|_{\mathcal{P}}$ is a 4×4 matrix and thus with 16 freely selectable numbers. For $n = 3 : 3 \times 3$ numbers.

Result: Therefore, with this free selectable 16 numbers it is possible to configure the 10 numbers of $g'\Big|_{\mathcal{P}}$ as desired and to bring them into a canonical form. For $n=3$: 9 numbers can configure 6 numbers.

Second term.

(6.35): $\left(\partial_{\underbrace{k'}_{4}} \underbrace{g'}_{10}\right)\Big|_{\mathcal{P}}$ is an array of $4 \times 10 = 40$ numbers. For $n = 3 : 3 \times 6 = $ 18 numbers.

(6.36): Because of $\dfrac{\partial}{\partial x^{k'}}\dfrac{\partial x^i}{\partial x^{i'}} = \underbrace{\dfrac{\partial}{\partial x^{i'}}\overset{\overbrace{4}}{\dfrac{\partial x^i}{\partial x^{k'}}}}_{10}$, partial derivatives commutate,

$\left(\dfrac{\partial}{\partial x^{k'}}\left(\dfrac{\partial x^i}{\partial x^{i'}}\dfrac{\partial x^j}{\partial x^{j'}}g_{ij}\right)\right)\Big|_{\mathcal{P}}$ has exactly $4 \times 10 = 40$ freely selectable numbers. For $n = 3 : 3 \times 6 = 18$ numbers.

Result: And hence it is possible to determine all the first derivatives of the metric tensor and set them to zero. For $n = 3 : 18$ numbers can configure 18 numbers.

Third term.

(6.35): $\left(\underbrace{\partial_{l'}\partial_{k'}}_{10}\,\underbrace{g'}_{10}\right)\Big|_{\mathcal{P}}$ is an array of $10 \times 10 = 100$ numbers (partial derivatives commutate). For $n = 3 : 6 \times 6 = 36$ numbers.

(6.36): $\left(\underbrace{\dfrac{\partial}{\partial x^{l'}}\dfrac{\partial}{\partial x^{k'}}\overset{\overbrace{4}}{\dfrac{\partial x^i}{\partial x^{i'}}}}_{20}\dfrac{\partial x^j}{\partial x^{j'}}g_{ij}\right)\Big|_{\mathcal{P}}$ is an array of $4 \times 20 = 80$ numbers. The

20 variations of $\dfrac{\partial}{\partial x^{l'}}\dfrac{\partial}{\partial x^{k'}}\dfrac{\partial x^i}{\partial x^{i'}}$ can again be explained by commutativity of the partial derivatives. Combinatorics tells us that with a set of n elements and a selection of k elements (with repetition) there are $\dbinom{n + k - 1}{k}$ possibilities.

And thus: $\dbinom{4 + 3 - 1}{3} = 20$. For $n = 3 : 3 \times 10 = 30$ numbers.

Result: The second derivatives of the metric cannot be made to vanish completely.

Summary: The metric tensor $\mathbf{g}$ can be transformed into a canonical form (see Section 4.4), and the vector space receives a Euclidean structure up to order $\mathcal{O}\left((\Delta x^i)^2\right)$ at a point $\mathcal{P}$:

$$g_{ij}(\mathcal{P}) = \pm\delta_{ij}, \text{(orthonormal basis at } \mathcal{P}\text{)},$$

$$\left(\partial_{k'}g'\right)\Big|_{\mathcal{P}} = 0, \text{(orthonormal basis is a good approximation close at } \mathcal{P}\text{)},$$

$$\left(\partial_{l'}\partial_{k'}g'\right)\Big|_{\mathcal{P}} \neq 0, \text{(not necessarily all components zero)}.$$

Example 6.5 *canonical metric*

The coordinate basis of a polar coordinate system is, equation (1.9) and (1.10),

$$\partial_r = \cos\theta\,\partial_x + \sin\theta\,\partial_y,$$
$$\partial_\theta = -r\sin\theta\,\partial_x + r\cos\theta\,\partial_y,$$

with the associated metric tensor

$$(g_{ij}) = \begin{pmatrix} \partial_r \cdot \partial_r & \partial_r \cdot \partial_\theta \\ \partial_\theta \cdot \partial_r & \partial_\theta \cdot \partial_\theta \end{pmatrix} = \begin{pmatrix} 1 & 0 \\ 0 & r^2 \end{pmatrix}.$$

The corresponding orthonormal basis is

$$\hat{e}_r = \partial_r, \quad \hat{e}_\theta = \frac{1}{r}\partial_\theta.$$

And with this orthonormal basis the metric has a canonical form:

$$(g_{ij})_{\text{orthonormal basis}} = (g_{ij})_{\text{canonical}} = \begin{pmatrix} 1 & 0 \\ 0 & 1 \end{pmatrix}.$$

Chapter 7

Curved space

7.1 Manifold

The term *manifold* (german 'Mannigfaltigkeit') came up in the 19th century, when mathematicians, especially Hermann Grassmann (1809-1877), discussed whether there are spaces with more than three dimensions.

A *manifold M* is a space of continuously arranged points in which a physical system can be defined *without coordinates*. Physical relationships must be valid independent of the selection of a coordinate system (covariance of physical equations). Such a points-space can be curved and of any complicated topology. An important condition of such an n-dimensional manifold is that each point contains an open neighbourhood which can be mapped homeomorphic, one-to-one to an open set of the $\mathbb{R}^n$. One speaks then that the manifold behaves locally like a Euclidean space $\mathbb{R}^n$.

A chart ϕ maps an open set U of the manifold M one-to-one to an image $\phi(U)$ of the $\mathbb{R}^n$, where $n = \dim M$ (see Figure 7.1).

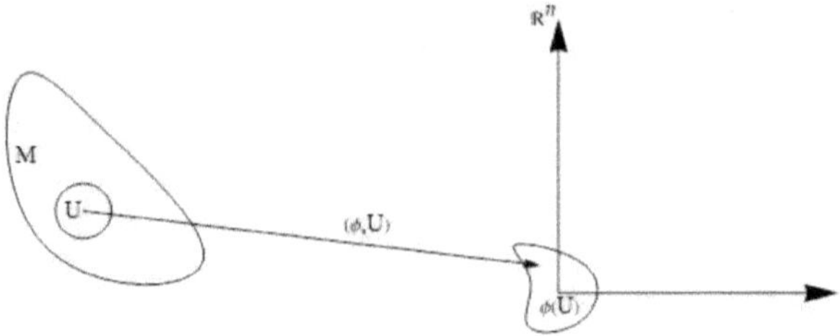

Figure 7.1: A chart ϕ maps an open set U of the manifold M.

This chart method can be applied to the entire manifold M: Cover the manifold M with open subsets, so that every point of M lies in at least one such subset. Then perform the one-to-one mapping to obtain a set of charts $\{\phi, \psi, \ldots\}$ of M. If two subsets in M partially overlap, their images on $\mathbb{R}$ also partially overlap. The overlapping areas in $\mathbb{R}$ are connected by the functions $\left(\phi \circ \psi^{-1}\right)$ and $\left(\psi \circ \phi^{-1}\right)$ diffeomorphically (Appendix A "mathematical terms"), see Figure 7.2. The smooth and overlapping sequence of such charts

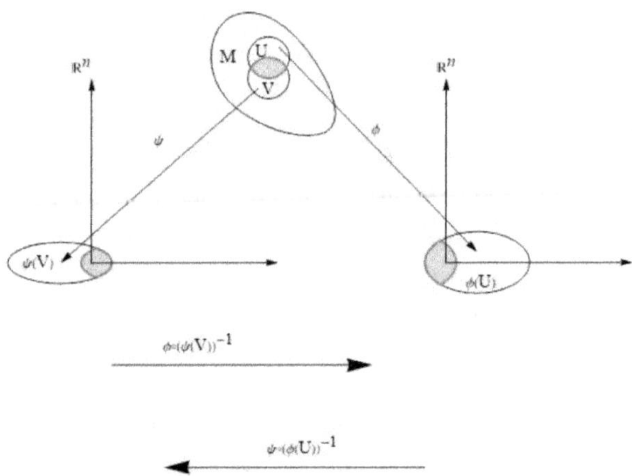

Figure 7.2: The overlap area on M is mapped to the grey coloured areas.

is called an atlas. Such an atlas clearly represents the manifold M. In this way one has a faithful image of the whole manifold, however complicated it may be in global terms. Thus it is possible to specify physical relations and laws on non-Euclidean spaces.

Curves are of high importance to determine and describe geometric objects (e.g. vector fields) and physical processes in a manifold. For example, the solutions of equations of motion are curves in the manifold. A curve on manifold M is a continuous series of points in M. By means of a complete atlas of M one can follow a curve as a whole from chart to chart. *The differentiability of such a curve is, as always, only given by charts in Euclidean spaces.* Figure 7.2 shows the following situation: $\phi(U)$ is the chart with coordinates $\mathbb{R}^n \left(x^i\right)$, and $\psi(V)$ is the chart with $\mathbb{R}^n \left(y^j\right)$. And so in the *overlap* region of U and V the chain rule for partial derivatives can be applied:

$$\frac{\partial}{\partial x^i} \underbrace{\left(\psi \circ \phi^{-1}\right)}_{y^j\left(x^i\right)} = \frac{\partial \left(\psi \circ \phi^{-1}\right)}{\partial y^j} \frac{\partial y^j}{\partial x^i}. \tag{7.1}$$

7.2 Intrinsic vectors

In a manifold we must abandon the conventional definition of a vector. The following example may illustrate this: Consider a curved surface (e.g. the surface of a sphere) on which 2-dimensional creatures live. It is impossible for them to determine a position vector that indicates a position of the curved surface from a location outside their physical sphere. Vectors must therefore be developed in a different way, namely by using *intrinsic geometry* (no embedding of the manifold in higher dimensional spaces). One defines intrinsic vectors along a curve: the *tangential vector*. We will derive this in the following.

Remark 7.1
Note that a manifold need not have a distance rule between points. Therefore a vector of a manifold cannot be defined by displacements Δx^i.

■

A curve $\mathcal{C}$ in M is a continuous series of points. We want to choose a different representation of this curve $\mathcal{C}$. Each point $\mathcal{P}$ of this curve is mapped with a value λ by means of a mapping g from $\mathbb{R}^1$ to M:

$$g : \mathbb{R}^1 \to M.$$

The point $\lambda \in \mathbb{R}^1$ is mapped to a point $\mathcal{C}$ in M. In this way the curve $\mathcal{C}$ is parametrized by λ, see Figure 7.3. We now have parametrized a curve $\mathcal{C}$ on

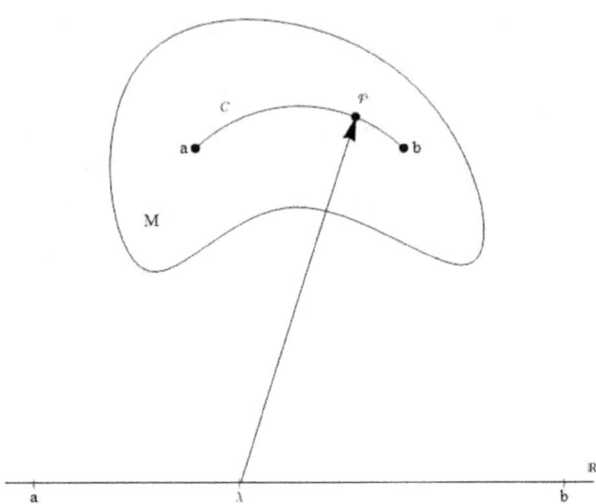

Figure 7.3: The point λ is mapped to the point $\mathcal{P}$ in M.

a manifold M with its points $\mathcal{P}(\lambda)$. On the other hand, we have learned that the chart method can be used to assign coordinates from $\mathbb{R}^n$ to each point $\mathcal{P}$ by a chart $\phi\left(\mathcal{P}(x^i)\right)$, Section 7.1, Figure 7.1.

A differentiable and arbitrary function f, which assigns a value to each point of M, map the point $\mathcal{P}$ to $\mathbb{R}^1$: $f(\mathcal{P} \to \mathbb{R}^1)$. The combined map $f \circ \phi^{-1} \equiv f(x^i(\lambda))$ therefore provides a map from $\mathbb{R}^n$ to $\mathbb{R}^1$. After differentiation with respect to λ and application of the chain rule at $\mathcal{P}$ we obtain:

$$\frac{df}{d\lambda} = \frac{dx^i}{d\lambda}\frac{\partial f}{\partial x^i}.$$ (7.2)

We can give two interpretations to equation (7.2). First, $\frac{d}{d\lambda}$ is a *directional derivative operator* that assigns a value to a function f at $\mathcal{P}$:

$$\frac{d}{d\lambda}(f)\Big|_{\mathcal{P}} \longrightarrow \mathbb{R}^1.$$ (7.3)

And second, as equation (7.2) is true for any f, then $\frac{d}{d\lambda} = \frac{dx^i}{d\lambda}\frac{\partial}{\partial x^i}$. $\frac{d}{d\lambda}$ is a *tangent vector* $\vec{v}_{\mathcal{P}}$ of a curve $\mathcal{C}$ at point $\mathcal{P}$. Thus tangential vectors $\vec{v}_{\mathcal{P}}$ also act on smooth functions and cause their directional derivative:

$$\vec{v}_{\mathcal{P}}(f)\Big|_{\mathcal{P}} = v^i \frac{\partial f}{\partial x^i}\Big|_{\mathcal{P}} \longrightarrow \mathbb{R}^1,$$

$$\vec{v}_{\mathcal{P}} = \frac{d}{d\lambda} = \frac{dx^i}{d\lambda}\frac{\partial}{\partial x^i}.$$ (7.4)

$$\frac{d}{d\lambda}\Big|_{\mathcal{P}} \equiv \text{Tangentialvector}\left(\text{component } \frac{dx^i}{d\lambda}, \text{coordinate basis } \frac{\partial}{\partial x^i}\right).$$

All directional derivatives $\frac{d}{d\lambda}$ along curves at point $\mathcal{P}$ as well as all tangential vectors $\frac{d}{d\lambda} = \frac{dx^i}{d\lambda}\frac{\partial}{\partial x^i} = \vec{v}_\mathcal{P}$ at point $\mathcal{P}$ ("are two sides of a coin") form the same vector space (see Box 7.2), namely the **tangential space** $\mathcal{T}_\mathcal{P}$ at point $\mathcal{P}$.

Box 7.1   *tangential space $\mathcal{T}_\mathcal{P}$ at $\mathcal{P}$ on a manifold M*

At each point $\mathcal{P}$ of a smooth, but otherwise arbitrary manifold M, a vector space $\mathcal{T}_\mathcal{P}$ is attached, in which the tangential vectors $\vec{v}_\mathcal{P} \in \mathcal{T}_\mathcal{P}$ are at point $\mathcal{P}$. The tangential vector space $\mathcal{T}_\mathcal{P}$ has the same dimension n as M. If (ϕ, U) is a chart on M, which contains the point $\mathcal{P}$, the n vectors $\frac{\partial}{\partial x^i}$ form a basis of $\mathcal{T}_\mathcal{P}$.

The calculus operations of a manifold M are performed in the tangential space $\mathcal{T}_\mathcal{P}$, which is based at the point $\mathcal{P}$ of the manifold M.

Only tangential vectors at the same point $\mathcal{P}$ can be added! Vectors at different points have no relation to each other. The vectors are located in the tangential space $\mathcal{T}_\mathcal{P}$ and *not* in the manifold M.

The basis $\left\{\frac{\partial}{\partial x^i}\right\}$ of a chart $\phi(U)$ can be combined with the basis $\left\{\frac{\partial}{\partial y^j}\right\}$ of a chart $\psi(V)$ on the complete manifold M, (see equation (7.1)):

$$\frac{\partial}{\partial x^i} = \frac{\partial y^j}{\partial x^i}\frac{\partial}{\partial y^j}.$$

Box 7.2

A: The directional derivatives $\frac{d}{d\lambda}$ form a vector space.
B: The vectors $\frac{d}{d\lambda} = \frac{dx^i}{d\lambda}\frac{\partial}{\partial x^i}$ form a vector space.

A. Directional derivatives at $\mathcal{P}$:
Let f and g be two smooth functions on M at $\mathcal{P}$, a and b be two real numbers, so it holds:

Linearity:

$$\frac{d}{d\lambda}\left(af + bg\right) = a\frac{df}{d\lambda} + b\frac{dg}{d\lambda},$$

$$\left(a\frac{d}{d\lambda} + b\frac{d}{d\mu}\right) f = a\frac{df}{d\lambda} + b\frac{df}{d\mu}.$$

Leibniz rule:

$$\frac{d}{d\lambda}\left(f \cdot g\right) = \frac{df}{d\lambda}g + f\frac{dg}{d\lambda},$$

$$\left(a\frac{d}{d\lambda} + b\frac{d}{d\mu}\right)\left(f \cdot g\right) = a\frac{df}{d\lambda}g + af\frac{dg}{d\lambda} + b\frac{df}{d\mu}g + bf\frac{dg}{d\mu}$$

$$= \left(a\frac{df}{d\lambda} + b\frac{df}{d\mu}\right)g + f\left(a\frac{dg}{d\lambda} + b\frac{dg}{d\mu}\right).$$

B. Tangential vectors at $\mathcal{P}$:

We consider two curves through $\mathcal{P}$, which are parametrized with λ and μ:
$\frac{d}{d\lambda} = \frac{dx^i}{d\lambda}\frac{\partial}{\partial x^i}$, $\frac{d}{d\mu} = \frac{dx^i}{d\mu}\frac{\partial}{\partial x^i}$, and hence:

$$a\frac{d}{d\lambda} + b\frac{d}{d\mu} = \left(a\frac{dx^i}{d\lambda} + b\frac{dx^i}{d\mu}\right)\frac{\partial}{\partial x^i}, \quad a, b \in \mathbb{R}.$$

$\left(a\frac{dx^i}{d\lambda} + b\frac{dx^i}{d\mu}\right)$ are the components of a new vector, which in turn is a tangent vector of some curve through $\mathcal{P}$. Thus there will be a curve with the parameter ν, so that at $\mathcal{P}$ applies:

$$\frac{d}{d\nu} = \left(a\frac{dx^i}{d\lambda} + b\frac{dx^i}{d\mu}\right)\frac{\partial}{\partial x^i},$$

and with this

$$a\frac{d}{d\lambda} + b\frac{d}{d\mu} = \frac{d}{d\nu}.$$

The vector space complementary to the tangent space $\mathcal{T}_\mathcal{P}$ is the co-tangential space $\mathcal{T}_\mathcal{P}^\star$. To each point $\mathcal{P}$ of M a one-form $\tilde{p}_\mathcal{P}$ from $\mathcal{T}_\mathcal{P}^\star$ is assigned. And with equation (2.1) applies:

$$\tilde{p}_\mathcal{P}\left(\vec{v}_\mathcal{P}\right) \to \mathbb{R}. \tag{7.5}$$

The vectors of tangential space $\mathcal{T}_P$ and covectors (one-forms) of $\mathcal{T}_P^\star$ exist only at point $\mathcal{P}$, and therefore have no "extension" (*intrinsic*), although we will still draw them as an arrow with amount and direction for the sake of clarity. At point $\mathcal{P}$ we have a so-called *local inertial frame* (see Section 6.6, 'local flatness theorem'), where the following applies:

$$g_{ij(\mathcal{P})} = \pm\delta_{ij} \quad \text{(flat space metric)},$$
$$\partial_r g_{ij(\mathcal{P})} = 0, \Longrightarrow \Gamma^k_{ji} = 0, \tag{7.6}$$
$$\partial_s \partial_r g_{ij(\mathcal{P})} \neq 0, \Longrightarrow \partial_s \Gamma^k_{ji} \neq 0.$$

7.3   Vector fields on manifolds

A vector field is a vividly known and familiar term. Examples are velocity fields of particles, force fields (e.g. gravitation). A vector field on a manifold $\vec{v}_M$ is related to the tangential vectors $\vec{v}_p$ discussed in Section 7.2 in the following way: A vector field $\vec{v}_M$ selects from each point $\mathcal{P}$ of a manifold M a single tangential vector $\vec{v}_p$ of the tangent space $\mathcal{T}_P$. For example, given the stationary flow of gas particles, (equivalent to the vector field $\vec{v}_M$), in a specified space, (equivalent to the manifold M), the velocity distribution of the gas particles is uniquely defined at each point $\mathcal{P}$ of the specified space. The associated velocity vector, (corresponds to the tangential vector $\vec{v}_p$ selected by the vector field $\vec{v}_M$), lies in the tangential space related to this point. The vector field of a manifold is, expressed plainly, the "sum" of the *selected* tangential vectors. The vector field $\vec{v}_M$ is therefore located in the tangential spaces $\mathcal{T}_P$'s of the manifold.

According to Section 7.2, equation (7.4), a tangent vector $\vec{v}_p$ causes a directional derivative to smooth functions f on M. In the same way, a vector field $\vec{v}_M$ does the same:

$$\vec{v}_M(f)\Big|_M \longrightarrow \mathbb{R}^1. \tag{7.7}$$

A vector field $\vec{v}_M$ is smooth or differentiable if $\vec{v}_M(f)$ is smooth for all smooth functions f on M.

7.4 Riemann manifold

An amorphous, continuous accumulation of points is a structureless space. Only if you install a metric tensor in it does the amorphous accumulation of points acquire a structure and become measurable. A metric tensor of the form $g_{ij} = \delta_{ij}$ transforms the amorphous accumulation of points into an Euclidean space. Or, the metric tensor $(\mathbf{g}) = \begin{pmatrix} r^2 & 0 \\ 0 & r^2 \sin^2 \theta \end{pmatrix}$ transforms it into a spherical surface (see Box 7.3).

A differentiable manifold on which a symmetrical, positive-definite $(0, 2)$ tensor field $\mathbf{g}$, $\mathbf{g}(\vec{v}, \vec{v}) > 0$, acts is called a *Riemannian manifold*. The space of special and general relativity is called a pseudo-Riemannian.

Box 7.3

We want to calculate the metric tensor of a spherical surface with the coordinates (θ, ϕ). We use the spherical coordinates (r, θ, ϕ), with r as distance parameter from the origin, θ the latitude angle and ϕ the longitude angle. Therefore:

$$x = r \sin \theta \cos \phi, \ y = r \sin \theta \cos \phi, \ z = r \cos \theta,$$

$$\vec{e}_\theta = \frac{\partial}{\partial \theta} = \frac{\partial x}{\partial \theta}\frac{\partial}{\partial x} + \frac{\partial y}{\partial \theta}\frac{\partial}{\partial y} + \frac{\partial z}{\partial \theta}\frac{\partial}{\partial z}$$

$$= r \left(\cos \theta \, \vec{e}_x + \cos \theta \cos \phi \, \vec{e}_y - \sin \theta \, \vec{e}_z \right),$$

$$\vec{e}_\phi = \frac{\partial}{\partial \phi} = \frac{\partial x}{\partial \phi}\frac{\partial}{\partial x} + \frac{\partial y}{\partial \phi}\frac{\partial}{\partial y} + \frac{\partial z}{\partial \phi}\frac{\partial}{\partial z}$$

$$= r \left(- \sin \theta \sin \phi \, \vec{e}_x + \sin \theta \cos \phi \, \vec{e}_y \right),$$

$$g_{ij} = \vec{e}_i \cdot \vec{e}_j, \Rightarrow g_{\theta\theta} = r^2, g_{\theta\phi} = g_{\phi\theta} = 0, g_{\phi\phi} = r^2 \sin^2 \theta,$$

$$(g)_{\text{spherical surface}} = \begin{pmatrix} r^2 & 0 \\ 0 & r^2 \sin^2 \theta \end{pmatrix}.$$

7.5 Parallel-transport of vectors

Let us imagine a triangular figure with vertices A, B and C. We define a route $\mathcal{R}$ to get back to A from A via vertices B and C. Now we move a vector $\vec{v}$ along this route $\mathcal{R}$ parallel to itself. If at infinitely close points of the curve the vector remains parallel to itself and of the same length, one speaks of a *parallel-transport* of $\vec{v}$ along the curve. What do we see when we compare the direction of the original vector $\vec{v}$ and the vector finally $\vec{v}\,'$ at A? In a flat Euclidean space, $\vec{v}$ and $\vec{v}\,'$ are parallel. However, in a curved space, for example a spherical surface, the vectors $\vec{v}$ and $\vec{v}\,'$ enclose an angle. So are no longer parallel. The consequence is that *no* parallel vector fields can be defined globally on a curved manifold.

We want to derive a mathematical description for the condition that a vector $\vec{v}$ is parallel-transported along a curve $x^j(\lambda)$, parametrized by λ. The components v^i of the vector $\vec{v}$ must be constant along the curve. Therefore:

$$\frac{dv^i}{d\lambda} = 0. \tag{7.8}$$

And with equation (7.4) we get the result for a parallel-transport of a vector $\vec{v}$ in the direction of a vector $\vec{u}$ (here the tangent vector), taking into account the fact that if a tensor equation is valid in one coordinate system, it is also valid in all other coordinate systems:

$$\begin{aligned}
\frac{dv^i}{d\lambda} &= 0, \\
\frac{dv^i}{d\lambda} &= \left(\frac{dx^j}{d\lambda} \frac{\partial}{\partial x^j} \right) v^i = \left(u^j \frac{\partial}{\partial x^j} \right) v^i \\
&= \underbrace{u^j \partial_j v^i}_{\text{Cartesian basis}} = \underbrace{u^j \nabla_j v^i}_{\text{general basis}} = \nabla_{u^j \partial_j} v^i = \nabla_{\vec{u}} \vec{v}, \\
\nabla_{\vec{u}} \vec{v} &= \vec{0}.
\end{aligned} \tag{7.9}$$

From the equations (6.6) and (6.7), under the condition of a parallel transport along the coordinate x^j, the following important relationship results for the further calculations:

$$\begin{aligned}
\nabla_{\partial_j} \vec{v} &= \vec{e}_k \left(\partial_j v^k + \Gamma^k_{ji} v^i \right) = \vec{0}, \\
\nabla_{\partial_j} v^k &= \partial_j v^k + \Gamma^k_{ji} v^i = 0, \\
\partial_j v^k &= -\Gamma^k_{ji} v^i.
\end{aligned} \tag{7.10}$$

Thus we have found another description for the affine connection coefficient Γ^k_{ji}: It " rules " the parallel transport.

When two vectors are *parallel-transported*, their inner product remains constant (*metric compatibility*).

$$\nabla_i \left(\vec{a} \cdot \vec{b} \right) \; = \; \nabla_i \left(g_{jk} a^j b^k \right)$$

$$= \; \underbrace{\left(\nabla_i g_{jk} \right)}_{=0} a^j b^k + g_{jk} \underbrace{\left(\nabla_i a^j \right)}_{=0} b^k + g_{jk} a^j \underbrace{\left(\nabla_i b^k \right)}_{=0} = 0,$$

$$\nabla_i \left(\vec{a} \cdot \vec{b} \right) \; = \; 0. \tag{7.11}$$

The first term of the second line is identically zero because of equation (6.26) (covariant constancy of the metric), and the last two terms from constraint 7.10 of the parallel transport. Thus the magnitude of a vector and the included angle of two vectors remain constant during parallel-transport.

7.6   Geodesics

We want to define a *straight line* in Euclidean space as follows: It parallel-transports its own tangent. And we can also apply this definition to curved spaces: A *geodesic* parallel-transports the tangent vector $\vec{u}$ without changing its magnitude and direction. And therefore with equation (7.9):

$$\nabla_{\vec{u}} \vec{u} = \vec{0}. \tag{7.12}$$

With the equations (7.4) and (7.9) we transform, $\nabla_{\vec{u}} \vec{u} = \vec{0}$, into a component expression:

$$(\nabla_{\vec{u}} \vec{u})^j = u^i \left(\partial_i u^j + \Gamma^j_{ik} u^k \right)$$

$$= u^i \underbrace{\frac{\partial}{\partial x^i}}_{\frac{d}{d\lambda}} u^j + \Gamma^j_{ik} u^i u^k = \frac{du^j}{d\lambda} + \Gamma^j_{ik} u^i u^k$$

$$\overset{u^j = \frac{dx^j}{d\lambda}}{=} \frac{d^2 x^j}{d\lambda^2} + \Gamma^j_{ik} \frac{dx^i}{d\lambda} \frac{dx^k}{d\lambda}.$$

And thus we obtain the second order geodesic non-linear differential equation:

$$\frac{d^2x^j}{d\lambda^2} + \Gamma^j_{ik}\frac{dx^i}{d\lambda}\frac{dx^k}{d\lambda} = 0, \ j = 1, \ldots, n. \tag{7.13}$$

With the boundary conditions $x^i(\lambda_0)$ (initial position) and $u^i(\lambda_0)$ (initial direction) at $\lambda = \lambda_0$, the equation has a unique solution. Geometrically interpreted, geodesics describe the shortest connection between two points. Physically, it means the path of motion of a force-free particle.

Remark 7.2

If we change the parameter λ by a linear transformation $\kappa = a\lambda + b$, (a, b are constants), also gives a solution to equation (7.13). Thus λ is an *affine parameter* .

∎

Example 7.1

For the Euclidean plane, the geodesic differential equation (7.13) provides simple straight lines as solutions because the connection coefficients Γ^j_{ik} are all zero.

Geodesics on a sphere:

There are any number of geodesics on a sphere $\{\theta, \phi\}$. We are looking for a solution with the following condition, that the geodesic has only a ϕ-dependency (longitude) but none of (latitude): $\frac{d\phi}{d\lambda} = u$ (constant 'velocity') and $\frac{d\theta}{d\lambda} = 0$. The only connection coefficients of a sphere that are not zero are (see Example 7.2):

$$\Gamma^\theta_{\phi\phi} = -\sin\theta\,\cos\theta, \Gamma^\phi_{\theta\phi} = \Gamma^\phi_{\phi\theta} = \frac{\cos\theta}{\sin\theta}.$$

If we include the specifications made into the geodesic equation 7.13 we get on the one hand:

$$\frac{d^2\phi}{d\lambda^2} + \Gamma^\phi_{\theta\phi}\frac{d\theta}{d\lambda}\frac{d\phi}{d\lambda} + \Gamma^\phi_{\phi\theta}\frac{d\phi}{d\lambda}\frac{d\theta}{d\lambda} = 0 + \Gamma^\phi_{\theta\phi}\cdot 0 \cdot u + \Gamma^\phi_{\theta\phi}\cdot u \cdot 0 = 0.$$

This equation leads us to no answer. And on the other hand:

$$\frac{d^2\theta}{d\lambda^2} + \Gamma^\theta_{\phi\phi}\frac{d\phi}{d\phi}\frac{d\phi}{d\lambda} = 0 - \sin\theta\,\cos\theta \cdot u^2 = 0.$$

For the definition domain $\theta : \{0, \pi\}$ there are three solutions for this equation:

$$\theta_1 = 0, \ \theta_2 = \pi/2 \text{ and } \theta_3 = \pi.$$

θ_1 and $\theta_3 = \pi$ are only point solutions at the poles. θ_2 is the equator line.

∎

Box 7.4

In Section 4.5 we have determined the path length along a curve using the tangent vector of the curve, equation (4.23). If the curve is a geodesic, what result do we get?

On a geodesic the tangent vector $\vec{u}$ is parallel-transported and thus the scalar product remains constant, equation (7.10). And with (4.23) we obtain:

$$l(\lambda) = \int_{\lambda_0}^{\lambda} \underbrace{\left|\sqrt{\vec{u}\cdot\vec{u}}\right|}_{\text{constant on geodesic}} d\lambda = \left|\sqrt{\vec{u}\cdot\vec{u}}\right| \int_{\lambda_0}^{\lambda} d\lambda$$

$$= \left|\sqrt{\vec{u}\cdot\vec{u}}\right| \lambda - \left|\sqrt{\vec{u}\cdot\vec{u}}\right| \lambda_0 = a\,\lambda + b.$$

We get the "surprising" result that the length calculation on a geodesic is identical to that of a line segment on a *straight line* in flat Euclidean space. And a second: the *length* on a geodesic is an *affine parameter* .

7.7   The curvature tensor

As we have already explained in Section 7.5, the parallel transport of a vector via a closed path (*holonomy*) is an indication of whether the space is flat or curved. The loop can have an infinitesimal extent, so that a local statement about the curvature of the space is obtained: the *Riemann curvature tensor*.

On a manifold we imagine an infinitesimal area with the coordinate lines x^1 and x^2, see Figure 7.4. We parallel-transport a vector from A via B to C, and

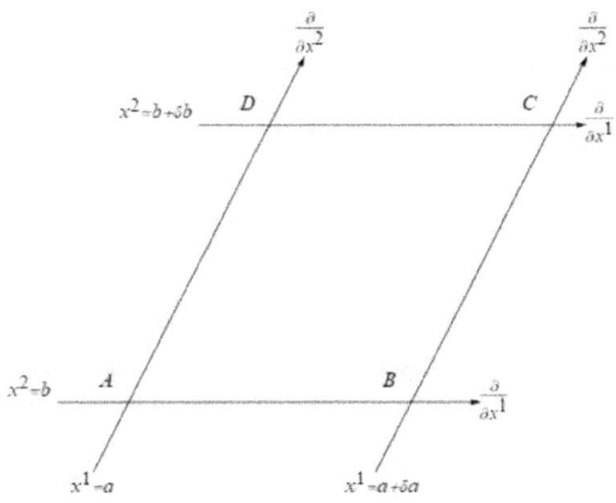

Figure 7.4: x^1-x^2 coordinate grid.

in the second path from A via D to C and compare the results. You get the same result if you integrate $ADCBA$ via the loop. Here we have chosen the difference between two paths, because it best reflects the commutator $\left[\frac{\partial}{\partial x^2}, \frac{\partial}{\partial x^1}\right]$ of the paths (see Section 7.11). We'll call it: commutator of a parallel transport.

A parallel transport of a vector $\vec{v}$ along a coordinate x^i has the following formula for the component v^j (see equation (7.10)):

$$\nabla_i \vec{v} = \vec{0},$$
$$j - \text{component: } \nabla_i v^j = \partial_i v^j + \Gamma^j_{ik} v^k = 0,$$
$$\partial_i v^j = \frac{\partial v^j}{\partial x^i} = -\Gamma^j_{ik} v^k.$$

By means of the integration of the expression $\frac{\partial v^j}{\partial x^i}$ along the coordinate x^i we get the change of the component v^j along the integration path $\int dx^i$.

Path 1:

$$\vec{v}(A) \longrightarrow \vec{v}(B): \ v^j(B) - v^j(A) = \underbrace{\int_{x^1=a}^{x^1=a+\delta a} \frac{\partial v^j}{\partial x^1} dx^1}_{x^2=b} = -\underbrace{\int_{x^1=a}^{x^1=a+\delta a} \Gamma^j_{1k} v^k dx^1}_{x^2=b},$$

$$\vec{v}(B) \longrightarrow \vec{v}(C): \ v^j(C) - v^j(B) = \underbrace{\int_{x^2=b}^{x^2=b+\delta b} \frac{\partial v^j}{\partial x^2} dx^2}_{x^1=a+\delta a} = -\underbrace{\int_{x^2=b}^{x^2=b+\delta b} \Gamma^j_{2k} v^k dx^2}_{x^1=a+\delta a},$$

$$\Longrightarrow$$

$$\left(v^j(B) - v^j(A) \right) + \left(v^j(C) - v^j(B) \right) =$$

$$v^j_{\text{path 1}} = -\underbrace{\int_{x^1=a}^{x^1=a+\delta a} \Gamma^j_{1k} v^k dx^1}_{x^2=b} - \underbrace{\int_{x^2=b}^{x^2=b+\delta b} \Gamma^j_{2k} v^k dx^2}_{x^1=a+\delta a}.$$

Path 2:

$$\vec{v}(A) \longrightarrow \vec{v}(D): \ v^j(D) - v^j(A) = \underbrace{\int_{x^2=b}^{x^2=b+\delta b} \frac{\partial v^j}{\partial x^2} dx^2}_{x^1=a} = -\underbrace{\int_{x^2=b}^{x^2=b+\delta b} \Gamma^j_{2k} v^k dx^2}_{x^1=a},$$

$$\vec{v}(D) \longrightarrow \vec{v}(C): \ v^j(C) - v^j(D) = \underbrace{\int_{x^1=a}^{x^1=a+\delta a} \frac{\partial v^j}{\partial x^1} dx^1}_{x^2=b+\delta b} = -\underbrace{\int_{x^1=a}^{x^1=a+\delta a} \Gamma^j_{1k} v^k dx^1}_{x^2=b+\delta b},$$

$$\Longrightarrow$$

$$\left(v^j(D) - v^j(A) \right) + \left(v^j(C) - v^j(D) \right) =$$

$$v^j_{\text{path 2}} = -\underbrace{\int_{x^2=b}^{x^2=b+\delta b} \Gamma^j_{2k} v^k dx^2}_{x^1=a} - \underbrace{\int_{x^1=a}^{x^1=a+\delta a} \Gamma^j_{1k} v^k dx^1}_{x^2=b+\delta b}.$$

The net change of $v^j(C)$ over the paths is thus

$$\delta v^j = v^j_{\text{path 2}} - v^j_{\text{path 1}}$$

$$= \left(-\underbrace{\int_{x^2=b}^{x^2=b+\delta b} \Gamma^j_{2k} v^k dx^2}_{x^1=a} - \underbrace{\int_{x^1=a}^{x^1=a+\delta a} \Gamma^j_{1k} v^k dx^1}_{x^2=b+\delta b} \right) -$$

$$- \left(-\underbrace{\int_{x^1=a}^{x^1=a+\delta a} \Gamma^j_{1k} v^k dx^1}_{x^2=b} - \underbrace{\int_{x^2=b}^{x^2=b+\delta b} \Gamma^j_{2k} v^k dx^2}_{x^1=a+\delta a} \right)$$

$$= \left(-\underbrace{\int_{x^2=b}^{x^2=b+\delta b} \Gamma^j_{2k} v^k dx^2}_{x^1=a} + \underbrace{\int_{x^2=b}^{x^2=b+\delta b} \Gamma^j_{2k} v^k dx^2}_{x^1=a+\delta a} \right) +$$

$$+ \left(\underbrace{\int_{x^1=a}^{x^1=a+\delta a} \Gamma^j_{1k} v^k dx^1}_{x^2=b} - \underbrace{\int_{x^1=a}^{x^1=a+\delta a} \Gamma^j_{1k} v^k dx^1}_{x^2=b+\delta b} \right).$$

We develop up to the first order of the Taylor series with regard to the co-ordinate grid spacing with infinitesimal δa and δb:

$$\delta v^j = v^j_{\text{path 2}} - v^j_{\text{path 1}}$$

$$= \left(-\underbrace{\int_{x^2=b}^{x^2=b+\delta b} \left(\Gamma^j_{2k} v^k + \delta a \frac{\partial}{\partial x^1}\left(\Gamma^j_{2k} v^k \right) + \cdots \right) dx^2}_{x^1 \to a+\delta a} + \underbrace{\int_{x^2=b}^{x^2=b+\delta b} \Gamma^j_{2k} v^k dx^2}_{x^1=a+\delta a} \right) +$$

$$+ \left(\underbrace{\int_{x^1=a}^{x^1=a+\delta a} \left(\Gamma^j_{1k} v^k + \delta b \frac{\partial}{\partial x^2}\left(\Gamma^j_{1k} v^k \right) + \cdots \right) dx^1}_{x^2 \to b+\delta b} - \underbrace{\int_{x^1=a}^{x^1=a+\delta a} \Gamma^j_{1k} v^k dx^1}_{x^2=b+\delta b} \right)$$

$$= -\int_{x^2=b}^{x^2=b+\delta b} \delta a \frac{\partial}{\partial x^1}\left(\Gamma^j_{2k} v^k \right) dx^2 + \int_{x^1=a}^{x^1=a+\delta a} \delta b \frac{\partial}{\partial x^2}\left(\Gamma^j_{1k} v^k \right) dx^1 + \mathcal{O}(\delta a^2, \delta b^2)$$

$$\approx \delta a \delta b \left(\frac{\partial}{\partial x^2}\left(\Gamma^j_{1k} v^k \right) - \frac{\partial}{\partial x^1}\left(\Gamma^j_{2k} v^k \right) \right).$$

In the transition to the last line, $\int_x^{x+\delta x} f(x)dx = f(x)\delta x$ was used for infinitesimal distances δx.

We will again make use of equation (7.10) , $\partial_j v^k = -\Gamma^k_{ji} v^i$, to completely reduce the obtained expression back to the connection coefficient Γ.

$$\delta v^j = \delta a \delta b \left(\frac{\partial}{\partial x^2} \left(\Gamma^j_{1k} v^k \right) - \frac{\partial}{\partial x^1} \left(\Gamma^j_{2k} v^k \right) \right)$$

$$= \delta a \delta b \left(\left(\partial_2 \Gamma^j_{1k} \right) v^k + \Gamma^j_{1k} \partial_2 v^k - \left(\partial_1 \Gamma^j_{2k} \right) v^k - \Gamma^j_{2k} \partial_1 v^k \right)$$

$$= \delta a \delta b \left(\left(\partial_2 \Gamma^j_{1k} \right) v^k - \underbrace{\Gamma^j_{1k} \Gamma^k_{2m} v^m}_{m \leftrightarrow k} - \left(\partial_1 \Gamma^j_{2k} \right) v^k + \underbrace{\Gamma^j_{2k} \Gamma^k_{1m} v^m}_{m \leftrightarrow k} \right)$$

$$= \delta a \delta b \underbrace{\left(\partial_2 \Gamma^j_{1k} - \Gamma^j_{1m} \Gamma^m_{2k} - \partial_1 \Gamma^j_{2k} + \Gamma^j_{2m} \Gamma^m_{1k} \right)}_{\text{so-called Riemann tensor}} v^k.$$

We modify the indices of the above result so that the components of the Riemann tensor $\mathbf{R}$ are comparable with other results in this textbook (Section 7.13): $j \longrightarrow k, k \longrightarrow s, 1\text{-coordinate} \longrightarrow i, 2\text{-coordinate} \longrightarrow j$.

$$
\begin{aligned}
(\delta v)^k &= R^k{}_{sji} \, (\delta v)^s \, (\delta x)^j \, (\delta x)^i , \\
R^k{}_{sji} &= \partial_j \Gamma^k_{is} - \partial_i \Gamma^k_{js} + \Gamma^k_{jl} \Gamma^l_{is} - \Gamma^k_{il} \Gamma^l_{js}.
\end{aligned}
\tag{7.14}
$$

Remark 7.3

The Riemann Tensor is a $(1,3)$ tensor with three slots for vectors $(\vec{v} \equiv v^s, \delta a \, \vec{e}_i, \delta b \, \vec{e}_j)$ and one slot for a one-form $\left(\tilde{e}^k, \tilde{e}^k \delta \vec{v} = \delta v^k \right)$. It reflects the *change in the components* δv^k of a vector v^s that is parallel-transported over a two-path route $(\delta a \, \vec{e}_i, \delta b \, \vec{e}_j)$ in reverse order. The Riemann tensor is only a function of the connection coefficients and finally of the metric, but not of the vector fields $\vec{v}, \vec{e}_i, \vec{e}_j$. And thus it represents a *local property* of the curved space.

■

7.8 Symmetry properties of the Riemann tensor

The properties of a tensor are coordinates independent. In order to investigate the *symmetry properties* of the Riemann tensor $\mathbf{R}$, we do so in a local inertial frame, Section 7.2, because there the expression is simplified due to $\Gamma^k_{ij} = 0$, conditions (7.6). And in this way we simplify the calculation work. With equation (7.14) we get:

$$R^k{}_{sji} = \partial_j \Gamma^k_{is} - \partial_i \Gamma^k_{js}.$$

We obtain the derivative of the connection coefficients Γ from the Levi-Civita connection formula, equation (6.25), noting that the derivative of the first factor g^{kt} is zero due to the constraints of a local inertial frame (7.6):

$$\partial_j \Gamma^k_{is} = \frac{1}{2} g^{kt} \left(\partial_j \partial_i g_{st} + \partial_j \partial_s g_{ti} - \partial_j \partial_t g_{is} \right),$$

$$\partial_i \Gamma^k_{js} = \frac{1}{2} g^{kt} \left(\partial_i \partial_j g_{st} + \partial_i \partial_s g_{tj} - \partial_i \partial_t g_{js} \right).$$

And therefore

$$\begin{aligned}
R^k{}_{sji} &= \partial_j \Gamma^k_{is} - \partial_i \Gamma^k_{js} \\
&= \frac{1}{2} g^{kt} \left(\partial_j \partial_i g_{st} + \partial_j \partial_s g_{ti} - \partial_j \partial_t g_{is} - \partial_i \partial_j g_{st} - \partial_i \partial_s g_{tj} + \partial_i \partial_t g_{js} \right).
\end{aligned}$$

Because of $\partial_d \partial_c g_{ab} = \partial_c \partial_d g_{ab}$ we get:

$$R^k{}_{sji} = \frac{1}{2} g^{kt} \left(\partial_j \partial_s g_{ti} - \partial_j \partial_t g_{is} - \partial_i \partial_s g_{tj} + \partial_i \partial_t g_{js} \right).$$

We multiply both sides by g_{lk}, $g_{lk} g^{kt} = \delta^t_l$, subsequent $l \longrightarrow k$, and obtain:

$$R_{ksji} = \frac{1}{2} \left(\partial_j \partial_s g_{ki} - \partial_j \partial_k g_{is} - \partial_i \partial_s g_{kj} + \partial_i \partial_k g_{js} \right). \tag{7.15}$$

Remark 7.4

The expression (7.15) for the $(0, 4)$ low-index Riemann tensor was derived assuming an inertial frame. It is used exclusively to investigate the symmetry properties of the Riemann tensor, because the symmetry properties of a tensor are independent of the coordinate system. To determine the components of the $(1, 3)$ Riemann tensor (7.14) an index raising by means of the metric is necessary. See the further explanations and Remark 7.5. ∎

The following four symmetry properties of the Riemann tensor can be read very clearly from equation (7.15):

$$\text{index 1-2 swap} : R_{ksji} = -R_{skji},$$
$$\text{index 3-4 swap} : R_{ksji} = -R_{ksij},$$
$$\text{index (12)-(34) swap} : R_{ksji} = R_{jiks}, \qquad (7.16)$$
$$\text{Bianchi identity} : R_{ksji} + R_{kjis} + R_{kisj} = 0.$$

Box 7.5

In a 2-dimensional space there are $2 \times 2 \times 2 \times 2 = 16$ possible different components of the Riemann tensor. However, due to symmetry properties, as we analyse, they will be reduced to 4 non-zero low index components, but reduced to only *one independent* component.

$$\text{1-2 swap} : R_{11xx} = -R_{11xx} = 0, \quad R_{22xx} = -R_{22xx} = 0,$$
$$\text{3-4 swap} : R_{xx11} = -R_{xx11} = 0, \quad R_{xx22} = -R_{xx22} = 0.$$

Four non-zero components remain

$$R_{1212} = a, \underbrace{\Longrightarrow}_{\text{3-4 swap}} R_{1221} = -a, \underbrace{\Longrightarrow}_{\text{1-2 swap}} R_{2121} = a, \underbrace{\Longrightarrow}_{\text{3-4 swap}} R_{2112} = -a,$$

with the scalar value $\pm a$.

Example 7.2

We want to determine the component $R^r{}_{\theta r \theta}$ of the Riemann tensor in polar coordinates for Euclidean plane. Of course we expect the components of $\mathbf{R}$ to be zero, because in a flat manifold, independent of the selected coordinate frame, a vector can be shifted parallel on any closed path and brought back to coincidence.

We use the results of Box 6.3 where we have already determined the connection coefficients for polar coordinates. With (7.14) we obtain:

$$R^r{}_{\theta r \theta} = \partial_r \Gamma^r_{\theta\theta} - \partial_\theta \Gamma^r_{r\theta} + \Gamma^r_{rl}\Gamma^l_{\theta\theta} - \Gamma^r_{\theta l}\Gamma^l_{r\theta}$$
$$= \partial_r(-r) - \partial_\theta(0) + \underbrace{\Gamma^r_{rr}}_{=0}\Gamma^r_{\theta\theta} + \underbrace{\Gamma^r_{r\theta}}_{=0}\Gamma^\theta_{\theta\theta} - \underbrace{\Gamma^r_{\theta r}}_{=0}\Gamma^r_{r\theta} - \underbrace{\Gamma^r_{\theta\theta}}_{=-r}\underbrace{\Gamma^\theta_{r\theta}}_{=\frac{1}{r}}$$
$$= -1 - (-1) = 0. \qquad \blacksquare$$

The issues of the low index components R_{ksji} of the Riemann tensor in comparison to $R^k{}_{sji}$, as pointed out in Remark 7.4 , is illustrated by the following for a 2-dimensional spherical surface $\{\theta, \phi\}$ with unit radius. First, the component $R^\theta{}_{\phi\theta\phi}$ is to be determined.

In Box 7.3 we have already calculated the associated metric data:

$$g_{\theta\theta} = 1, \; g_{\theta\phi} = g_{\phi\theta} = 0, \; g_{\phi\phi} = \sin^2\theta, \; r = 1.$$

With this data we find the connection coefficients, equation (6.25) :

$$\Gamma^k{}_{ji} = \frac{1}{2}g^{ks}\left(\partial_j\, g_{is} + \partial_i\, g_{sj} - \partial_s\, g_{ji}\right).$$

For the sake of clarity, we have written out the summation over s for the first expression. This is omitted in the determination of the other coefficients.

$$\Gamma^\theta{}_{\theta\theta} = \frac{1}{2}g^{\theta s}\left(\partial_\theta\, g_{\theta s} + \partial_\theta\, g_{s\theta} - \partial_s\, g_{\theta\theta}\right)$$

$$= \frac{1}{2}g^{\theta\theta}\left(\partial_\theta\, g_{\theta\theta} + \partial_\theta\, g_{\theta\theta} - \partial_\theta\, g_{\theta\theta}\right) + \frac{1}{2}g^{\theta\phi}\left(\partial_\theta\, g_{\theta\phi} + \partial_\theta\, g_{\phi\theta} - \partial_\phi\, g_{\theta\theta}\right) = 0,$$

$$\Gamma^\theta{}_{\theta\phi} = \frac{1}{2}g^{\theta s}\left(\partial_\theta\, g_{\phi s} + \partial_\phi\, g_{s\theta} - \partial_s\, g_{\theta\phi}\right) = 0, \; \Gamma^\theta{}_{\phi\theta} = 0,$$

$$\Gamma^\theta{}_{\phi\phi} = \frac{1}{2}g^{\theta s}\left(\partial_\phi\, g_{\phi s} + \partial_\phi\, g_{s\phi} - \partial_s\, g_{\phi\phi}\right) = \frac{1}{2}g^{\theta\theta}\left(-\partial_\theta g_{\phi\phi}\right) = -\sin\theta\cos\theta,$$

$$\Gamma^\phi{}_{\phi\phi} = \frac{1}{2}g^{\phi s}\left(\partial_\phi\, g_{\phi s} + \partial_\phi\, g_{s\phi} - \partial_s\, g_{\phi\phi}\right) = 0,$$

$$\Gamma^\phi{}_{\theta\phi} = \frac{1}{2}g^{\phi s}\left(\partial_\theta\, g_{\phi s} + \partial_\phi\, g_{s\theta} - \partial_s\, g_{\theta\phi}\right) = \frac{1}{2}g^{\phi\phi}\left(\partial_\theta g_{\phi\phi}\right)$$

$$= \frac{1}{2}(\sin^2\theta)^{-1}(2\sin\theta\cos\theta) = \frac{\cos\theta}{\sin\theta} = \Gamma^\phi{}_{\phi\theta},$$

$$\Gamma^\phi{}_{\theta\theta} = \frac{1}{2}g^{\phi s}\left(\partial_\theta\, g_{\theta s} + \partial_\theta\, g_{s\theta} - \partial_s\, g_{\theta\theta}\right) = 0.$$

In summary all non-zero 2-spherical Christoffel connections:

$$\Gamma^\theta{}_{\phi\phi} = -\sin\theta\cos\theta, \; \Gamma^\phi{}_{\theta\phi} = \Gamma^\phi{}_{\phi\theta} = \frac{\cos\theta}{\sin\theta}. \tag{7.17}$$

If we had set the radius parameter $r \neq 1$, i.e. $g_{\theta\theta} = r^2$, $g_{\phi\phi} = r^2\sin^2\theta$, we would get the same result for the connections, as can easily be calculated. We now have all the data together to determine $R^\theta{}_{\phi\theta\phi}$. With (7.14) we obtain:

$$
\begin{aligned}
R^\theta{}_{\phi\theta\phi} &= \partial_\theta \Gamma^\theta_{\phi\phi} - \partial_\phi \Gamma^\theta_{\theta\phi} + \Gamma^\theta_{\theta l}\Gamma^l_{\phi\phi} - \Gamma^\theta_{\phi l}\Gamma^l_{\theta\phi} \\
&= \partial_\theta(-\sin\theta\cos\theta) - \partial_\phi \frac{\cos\theta}{\sin\theta} + \left(\Gamma^\theta_{\theta\theta}\Gamma^\theta_{\phi\phi} + \Gamma^\theta_{\theta\phi}\Gamma^\phi_{\phi\phi}\right) - \left(\Gamma^\theta_{\phi\theta}\Gamma^\theta_{\theta\phi} + \Gamma^\theta_{\phi\phi}\Gamma^\phi_{\theta\phi}\right) \\
&= -\cos^2\theta + \sin^2\theta - 0 + (0+0) - \left(0 + (-\sin\theta\cos\theta)\frac{\cos\theta}{\sin\theta}\right) \\
&= \sin^2\theta.
\end{aligned}
\tag{7.18}
$$

Remark 7.5

The investigated symmetry properties of the Riemann tensor in this section and summarized in equation (7.16), show for the considered case of a sphere $\{\theta,\phi\}$ that the following is valid:

$$
R_{\theta\phi\theta\phi} \overset{\text{1--2 swap}}{=} -R_{\phi\theta\theta\phi} \overset{\text{3--4 swap}}{=} R_{\phi\theta\phi\theta}.
$$

It would be a big mistake to assume from the shown equality relation between $R_{\theta\phi\theta\phi}$ and $R_{\phi\theta\phi\theta}$ that $R^\theta{}_{\phi\theta\phi}$ is as well equal to $R^\phi{}_{\theta\phi\theta}$. IT IS NOT! It is important to note that $R^\theta{}_{\phi\theta\phi}$ and $R_{\theta\phi\theta\phi}$ are two different pairs of boots.

∎

To determine the remaining three Riemann components of a 2-sphere we can continue with equation (7.14). A more clever way is to take advantage of the fact that the low index components of **R** for a 2-dimensional manifold differ only by sign. And with the help of the metric, an index raising $R^k{}_{sji} = g^{kk}R_{ksji}$ can then be performed. The starting point is the already determined component $R^\theta{}_{\phi\theta\phi} = \sin^2\theta$, equation (7.18). The assigned low-index component of $R^\theta{}_{\phi\theta\phi}$ is:

$$
R_{\theta\phi\theta\phi} = g_{\theta\theta}R^\theta{}_{\phi\theta\phi} = 1\cdot\sin^2\theta = \sin^2\theta.
$$

The remaining Riemann components are calculated as such:

$$
R^\phi{}_{\theta\phi\theta} = g^{\phi\phi}R_{\phi\theta\phi\theta} \overset{\text{1--2 swap}}{=} g^{\phi\phi}(-R_{\theta\phi\phi\theta}) \overset{\text{3--4 swap}}{=} -g^{\phi\phi}(-R_{\theta\phi\theta\phi}) = \frac{1}{\sin^2\theta}\sin^2\theta = 1,
$$

$$
R^\theta{}_{\phi\phi\theta} = g^{\theta\theta}R_{\theta\phi\phi\theta} \overset{\text{3--4 swap}}{=} g^{\theta\theta}(-R_{\theta\phi\theta\phi}) = -1\cdot\sin^2\theta = -\sin^2\theta,
$$

$$
R^\phi{}_{\theta\theta\phi} = g^{\phi\phi}R_{\phi\theta\theta\phi} \overset{\text{1--2 swap}}{=} g^{\phi\phi}(-R_{\theta\phi\theta\phi}) = -\frac{1}{\sin^2\theta}\sin^2\theta = -1.
$$

Summary 7.1
The four components of the Riemann tensor $\mathbf{R}$ for a 2-sphere surface $\{\theta, \phi\}$ are:

$$R^{\theta}_{\ \phi\theta\phi} = \sin^2\theta,$$
$$R^{\theta}_{\ \phi\phi\theta} = -\sin^2\theta,$$
$$R^{\phi}_{\ \theta\phi\theta} = 1,$$
$$R^{\phi}_{\ \theta\theta\phi} = -1.$$

7.9 Integral curves

Let there be an arbitrary smooth vector field $\vec{v}$ on a manifold M. We select any point $\mathcal{P}$ of the vector field $\vec{v}$ and form the tangent vector there. The curve that is always tangent to the vector field, starting from point $\mathcal{P}$, is called an *integral curve* or *flow curve*. Thus each vector field is also defined by its congruence with the integral curves, its tangent field.

How can a flow curve be determined for a given vector field $\vec{v}$? With a coordinate system $\{x^i\}$, the components of the vector field $\vec{v}$ are $v^i\{x^i\}$. The tangential vector field is given by (see also equation (7.4)):

$$\frac{dx^i}{d\lambda} = v^i(x^i). \tag{7.19}$$

Equation (7.19) reflects a set of first-order *differential equations*. Each *integral curve* of a vector field is a *particular solution* of such a set of differential equations. All solutions yield the vector field itself. With the initial conditions at a point $\mathcal{P}(\lambda_0)$, the system of equations has a unique solution. For this reason, integral curves can never cross except at the origin ($v^i = 0$ for all i).

Example 7.3
Given a vector field $\vec{v} = 4\,\vec{e}_x + x\,\vec{e}_y = 4\,\frac{\partial}{\partial x} + x\,\frac{\partial}{\partial y}$. We want to determine the

flow curve at the starting point $\mathcal{P}(0,1)$ for $\lambda = 0$. With (7.19) we get:

$$\vec{v} = 4\,\frac{\partial}{\partial x} + x\,\frac{\partial}{\partial y} = \frac{dx}{d\lambda}\frac{\partial}{\partial x} + \frac{dy}{d\lambda}\frac{\partial}{\partial y},$$

$$\frac{dx}{d\lambda} = 4, \quad \frac{dy}{d\lambda} = x.$$

By integrating the second line, we get:

$$dx = 4\,d\lambda, \Longrightarrow x = 4\lambda + c,$$

$$dy = x\,d\lambda = (4\lambda + c)\,d\lambda = 4\lambda\,d\lambda + c\,d\lambda, \Longrightarrow y = 2\lambda^2 + c\lambda + d.$$

For $\lambda = 0$ we have chosen the starting point $\mathcal{P}(0,1)$, and thus we get the integration constants $c = 0$ and $d = 1$. Figure 7.5 shows the vector field $\vec{v} = 4\,\vec{e}_x + x\,\vec{e}_y$ and the corresponding flow curve through point $\mathcal{P}(0,1)$.

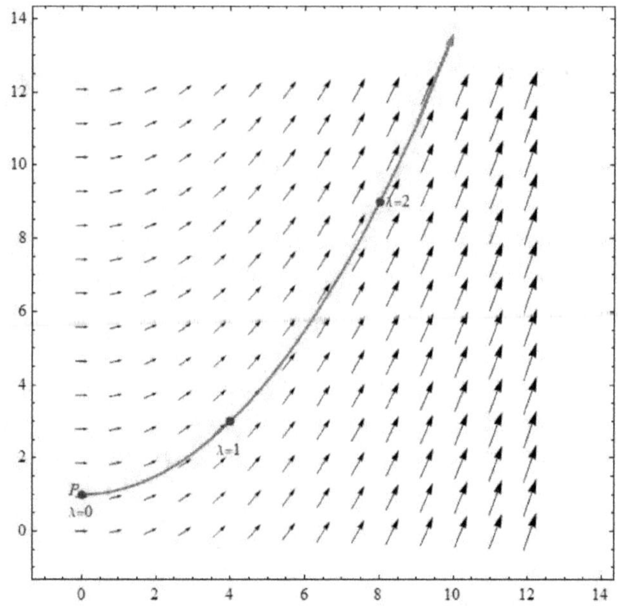

Figure 7.5: Vector field $\vec{v} = 4\,\vec{e}_x + x\,\vec{e}_y$ and flow curve through point $\mathcal{P}(0,1)$.

■

Example 7.4

A well-known example are the integral curves of the vector field

$$\vec{v} = -y\,\vec{e}_x + x\,\vec{e}_y.$$

With the approach from equation (7.19) we obtain

$$\frac{dx}{d\lambda} = -y, \frac{dy}{d\lambda} = x.$$

One can see that the trigonometric functions can help us here to find the following solution:

$$x = a \cos\lambda, \ y = a \sin\lambda, \ a \in \mathbb{R}.$$

This approach leads us directly to the basis vector $\vec{e}_\theta$ of polar coordinates, equation (1.10). The integral curves of $\vec{v}$ are therefore nothing else than the lines of motion of the vector $\vec{e}_\theta$, Figure 7.6.

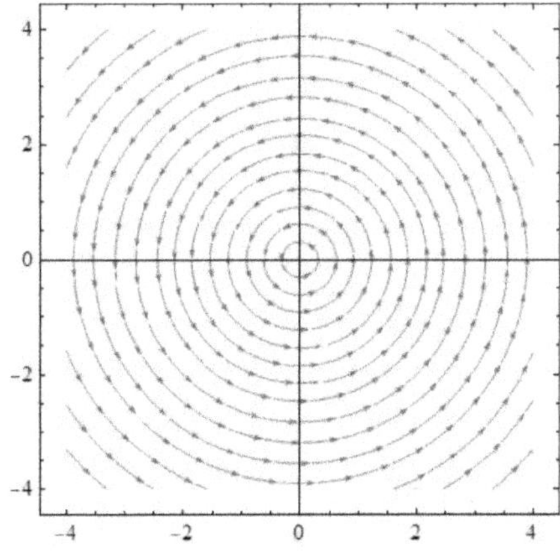

Figure 7.6: Integral curves of $\vec{v} = -y\,\vec{e}_x + x\,\vec{e}_y$.

■

Box 7.6

We want to show that the tangential vector $\vec{u}$ or its representation as the directional derivative $\frac{d}{d\lambda}$, equation (7.4), becomes a *translation operator* by exponentiation. We assume a manifold M and an integral curve $\vec{u} = \frac{d}{d\lambda}$ with the coordinates $x^i(\lambda)$. The coordinates of two points with the parameters λ_0 and $\lambda_0 + \delta$ are given by the Taylor series:

$$
\begin{aligned}
x^i(\lambda_0 + \delta) &= x^i(\lambda_0) + \delta \frac{dx^i}{d\lambda}\Big|_{\lambda_0} + \frac{1}{2!}\delta^2 \frac{d^2 x^i}{d\lambda^2}\Big|_{\lambda_0} + \frac{1}{3!}\delta^2 \frac{d^3 x^i}{d\lambda^3}\Big|_{\lambda_0} + \cdots \\
&= \left(1 + \delta \frac{d}{d\lambda} + \frac{1}{2!}\delta^2 \frac{d^2}{d\lambda^2} + \frac{1}{3!}\delta^3 \frac{d^3}{d\lambda^2} + \cdots \right) x^i\Big|_{\lambda_0} \\
&= \exp\left[\delta \frac{d}{d\lambda}\right] x^i\Big|_{\lambda_0} = e^{\delta \frac{d}{d\lambda}} x^i\Big|_{\lambda_0} = e^{\delta \vec{u}} x^i\Big|_{\lambda_0}.
\end{aligned}
\tag{7.20}
$$

$\delta \frac{d}{d\lambda}$ is an infinitesimal increment along the integral curve. A finite progress is achieved with the translation operator $e^{\delta \frac{d}{d\lambda}}$.

7.10   Non-coordinate bases

A coordinate system $\{x^i\}$ is characterized in that x^i is constant along the $x^j, j \neq i$, lines. A coordinate line x^i is the integral curve of $\frac{\partial}{\partial x^i}$. For this reason, $\frac{\partial}{\partial x^i}$ and $\frac{\partial}{\partial x^j}$ commute for all i, j, $\left[\frac{\partial}{\partial x^i}, \frac{\partial}{\partial x^j}\right] = 0$. Therefore the basis $\left\{\frac{\partial}{\partial x^i}\right\}$ is called *coordinate basis* (see Section 1.4).

The integral curves $\frac{d}{d\lambda}$ and $\frac{d}{d\mu}$ of two vector fields $\vec{u}$ and $\vec{v}$ generally do not form a coordinate basis, i.e. $\left[\frac{d}{d\lambda}, \frac{d}{d\mu}\right] \neq 0$. And that's why μ is not constant along $\frac{d}{d\lambda}$, in contrast to a coordinate basis. In this case one speaks of a *non-coordinate basis* .

Let us calculate the commutator of two vector fields $\vec{u} = \frac{d}{d\lambda}$ and $\vec{v} = \frac{d}{d\mu}$.

We use equations (7.4) and (7.19) and get:

$$
\begin{aligned}
\left[\frac{d}{d\lambda}, \frac{d}{d\mu}\right] &= \frac{d}{d\lambda}\frac{d}{d\mu} - \frac{d}{d\mu}\frac{d}{d\lambda} \\
&= \frac{dx^i}{d\lambda}\frac{\partial}{\partial x^i}\left(\frac{dx^j}{d\mu}\frac{\partial}{\partial x^j}\right) - \frac{dx^j}{d\mu}\frac{\partial}{\partial x^j}\left(\frac{dx^i}{d\lambda}\frac{\partial}{\partial x^i}\right) \\
&= u^i\frac{\partial}{\partial x^i}\left(v^j\frac{\partial}{\partial x^j}\right) - v^j\frac{\partial}{\partial x^j}\left(u^i\frac{\partial}{\partial x^i}\right) \\
&= u^i\frac{\partial v^j}{\partial x^i}\frac{\partial}{\partial x^j} + u^i v^j\frac{\partial}{\partial x^i}\frac{\partial}{\partial x^j} - v^j\frac{\partial u^i}{\partial x^j}\frac{\partial}{\partial x^i} - v^j u^i\frac{\partial}{\partial x^j}\frac{\partial}{\partial x^i} \\
&= u^i\frac{\partial v^j}{\partial x^i}\frac{\partial}{\partial x^j} - \underbrace{v^j\frac{\partial u^i}{\partial x^j}\frac{\partial}{\partial x^i}}_{i \leftrightarrow j} + u^i v^j\underbrace{\left(\frac{\partial}{\partial x^i}\frac{\partial}{\partial x^j} - \frac{\partial}{\partial x^j}\frac{\partial}{\partial x^i}\right)}_{=0,\ \text{coordinate basis}} \\
&= \underbrace{\left(u^i\frac{\partial v^j}{\partial x^i} - v^i\frac{\partial u^j}{\partial x^i}\right)\frac{\partial}{\partial x^j}}_{j-\text{component}}.
\end{aligned}
$$

The important result is that the commutator of two vector fields is again a vector field:

$$
\begin{aligned}
\left[\frac{d}{d\lambda}, \frac{d}{d\mu}\right] = [\vec{u}, \vec{v}] &= \left(u^i\frac{\partial v^j}{\partial x^i} - v^i\frac{\partial u^j}{\partial x^i}\right)\frac{\partial}{\partial x^j}, \\
j-\text{component}: [\vec{u}, \vec{v}]^j &= \left(u^i\frac{\partial v^j}{\partial x^i} - v^i\frac{\partial u^j}{\partial x^i}\right).
\end{aligned}
\tag{7.21}
$$

If the components do not become zero, then the integral curves of the two vector fields form a non-coordinate basis.

Example 7.5

We will show that the unit vectors of the polar coordinates $\{\hat{e}_r, \hat{e}_\theta\}$ are a *non-coordinate* basis.

$$
\hat{e}_r = \cos\theta\frac{\partial}{\partial x} + \sin\theta\frac{\partial}{\partial y}, \quad \hat{e}_\theta = -\sin\theta\frac{\partial}{\partial x} + \cos\theta\frac{\partial}{\partial y},
$$

$$
\cos\theta = \frac{x}{r} = \frac{x}{\sqrt{x^2 + y^2}}, \quad \sin\theta = \frac{y}{r} = \frac{y}{\sqrt{x^2 + y^2}}.
$$

With equation (7.21) we get:

$$[\hat{e}_r, \hat{e}_\theta]^x = \left(\cos\theta \frac{\partial(-\sin\theta)}{\partial x} - (-\sin\theta)\frac{\partial\cos\theta}{\partial x} \right) + \left(\sin\theta\frac{\partial(-\sin\theta)}{\partial y} - \cos\theta\frac{\partial\cos\theta}{\partial y} \right)$$

$$= \left(\frac{x}{r}\frac{\partial}{\partial x}\left(\frac{-y}{\sqrt{x^2+y^2}} \right) - \left(\frac{-y}{r} \right)\frac{\partial}{\partial x}\left(\frac{x}{\sqrt{x^2+y^2}} \right) \right) +$$

$$+ \left(\frac{y}{r}\frac{\partial}{\partial y}\left(\frac{-y}{\sqrt{x^2+y^2}} \right) - \left(\frac{x}{r} \right)\frac{\partial}{\partial y}\left(\frac{x}{\sqrt{x^2+y^2}} \right) \right)$$

$$= \left(\frac{x}{r}\frac{xy}{r^3} + \frac{y}{r}\frac{y^2}{r^3} \right) + \left(\frac{y}{r}\left(\frac{-x^2}{r^3} \right) - \frac{x}{r}\left(\frac{-xy}{r^3} \right) \right)$$

$$= \frac{y^3 + x^2 y}{r^4} = \frac{y}{r^2} = \frac{1}{r}\sin\theta.$$

$$[\hat{e}_r, \hat{e}_\theta]^y = \left(\cos\theta\frac{\partial\cos\theta}{\partial x} - (-\sin\theta)\frac{\partial\sin\theta}{\partial x} \right) + \left(\sin\theta\frac{\partial\cos\theta}{\partial y} - \cos\theta\frac{\partial\sin\theta}{\partial y} \right)$$

$$= \left(\frac{x}{r}\frac{\partial}{\partial x}\left(\frac{x}{\sqrt{x^2+y^2}} \right) + \left(\frac{y}{r} \right)\frac{\partial}{\partial x}\left(\frac{y}{\sqrt{x^2+y^2}} \right) \right) +$$

$$+ \left(\frac{y}{r}\frac{\partial}{\partial y}\left(\frac{x}{\sqrt{x^2+y^2}} \right) - \left(\frac{x}{r} \right)\frac{\partial}{\partial y}\left(\frac{y}{\sqrt{x^2+y^2}} \right) \right)$$

$$= \left(\frac{x}{r}\frac{y^2}{r^3} + \frac{y}{r}\left(\frac{-xy}{r^3} \right) \right) + \left(\frac{y}{r}\left(\frac{-xy}{r^3} \right) - \frac{x}{r}\left(\frac{x^2}{r^3} \right) \right)$$

$$= -\frac{x^2+y^2}{r^3}\frac{x}{r} = -\frac{1}{r}\cos\theta.$$

And therefore

$$[\hat{e}_r, \hat{e}_\theta] = \frac{1}{r}\sin\theta\,\vec{e}_x - \frac{1}{r}\cos\theta\,\vec{e}_y = -\frac{1}{r}(-\sin\theta\,\vec{e}_x + \cos\theta\,\vec{e}_y) = -\frac{1}{r}\hat{e}_\theta \neq \vec{0},$$

what was to be demonstrated.

∎

In contrast to a normalised basis for polar coordinates, see Example 7.5, the polar coordinates $\{ \frac{\partial}{\partial r}, \frac{\partial}{\partial \theta} \}$, see equations (1.9),(1.10), are a coordinate basis, $[\frac{\partial}{\partial r}, \frac{\partial}{\partial \theta}] = \vec{0}$. What should be briefly shown here. With $x = r\cos\theta, y = r\sin\theta$

and $\frac{\partial}{\partial r} = \frac{\partial x}{\partial r}\frac{\partial}{\partial x} + \frac{\partial y}{\partial r}\frac{\partial}{\partial y}$, $\frac{\partial}{\partial \theta} = \frac{\partial x}{\partial \theta}\frac{\partial}{\partial x} + \frac{\partial y}{\partial \theta}\frac{\partial}{\partial y}$, we get using (7.21):

$$\left[\frac{\partial}{\partial r}, \frac{\partial}{\partial \theta}\right]^x = [\vec{e}_r, \vec{e}_\theta]^x$$

$$= \left(\cos\theta\frac{\partial(-r\sin\theta)}{\partial x} - (-r\sin\theta)\frac{\partial\cos\theta}{\partial x}\right) +$$

$$+ \left(\sin\theta\frac{\partial(-r\sin\theta)}{\partial y} - r\cos\theta\frac{\partial\cos\theta}{\partial y}\right)$$

$$= \left(\frac{x}{r}\frac{\partial}{\partial x}(-y) - (-y)\frac{\partial}{\partial x}\left(\frac{x}{\sqrt{x^2+y^2}}\right)\right) +$$

$$+ \left(\frac{y}{r}\frac{\partial}{\partial y}(-y) - (-x)\frac{\partial}{\partial y}\left(\frac{x}{\sqrt{x^2+y^2}}\right)\right)$$

$$= \left(\frac{x}{r}\cdot 0 + y\frac{y^2}{r^3}\right) + \left(\frac{y}{r}(-1) - x\frac{-xy}{r^3}\right)$$

$$= \frac{y^3}{r^3} - \frac{y}{r} + \frac{x^2 y}{r^3} = \frac{y}{r}\left(\frac{y^2+x^2}{r}\right) - \frac{y}{r} = 0.$$

$$\left[\frac{\partial}{\partial r}, \frac{\partial}{\partial \theta}\right]^y = [\vec{e}_r, \vec{e}_\theta]^y$$

$$= \left(\cos\theta\frac{\partial(r\cos\theta)}{\partial x} - (-r\sin\theta)\frac{\partial\sin\theta}{\partial x}\right) +$$

$$+ \left(\sin\theta\frac{\partial(r\cos\theta)}{\partial y} - r\cos\theta\frac{\partial\sin\theta}{\partial y}\right)$$

$$= \left(\frac{x}{r}\frac{\partial}{\partial x}(x) + y\frac{\partial}{\partial x}\left(\frac{y}{\sqrt{x^2+y^2}}\right)\right) +$$

$$+ \left(\frac{y}{r}\frac{\partial}{\partial y}(x) - x\frac{\partial}{\partial y}\left(\frac{y}{\sqrt{x^2+y^2}}\right)\right)$$

$$= \left(\frac{x}{r}\cdot 1 + y\left(\frac{-xy}{r^3}\right)\right) + \left(\frac{y}{r}\cdot 0 - x\frac{x^2}{r^3}\right)$$

$$= \frac{x}{r}\left(1 - \frac{y^2+x^2}{r^2}\right) = 0.$$

And therefore

$$\left[\frac{\partial}{\partial r}, \frac{\partial}{\partial \theta}\right] = [\vec{e}_r, \vec{e}_\theta] = \vec{0}.$$

7.11   Lie bracket

The commutator $\left[\frac{d}{d\lambda}, \frac{d}{d\mu}\right]$ (see Section 7.10) of the vector fields $\vec{u} = \frac{d}{d\lambda}$ and $\vec{v} = \frac{d}{d\mu}$ is again a vector field (compare equation (7.21)) and thus a tensor and is called *Lie bracket* . Its geometric interpretation is a vector field that represents an unclosed parallelogram of different flow curve paths. See Figure 7.7.

To put this into a mathematical form, we make the following consideration:

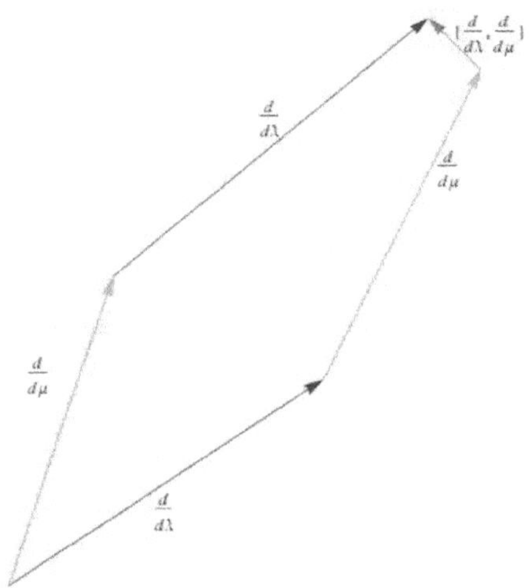

Figure 7.7: Simplified geometric interpretation of the Lie bracket, as a difference vector of different flow curves.

We choose a starting point P and go once along $\vec{u}$ for a distance $\Delta\lambda = \delta$, reach a point T, and then along $\vec{v}$ for the same distance $\Delta\mu = \delta$ and arrive at a point A. Now we only do the same thing in reverse: first along $\vec{v}$ and then along $\vec{u}$ by the same parameter difference δ, and get to a point B. We will now show that the vector from A to B is $\delta^2 [\vec{u}, \vec{v}]$ (lowest order of δ).

With equation (7.20) from Box 7.6 we get for the coordinates $x^i(T)$ and $x^i(A)$:

$$x^i(T) = e^{\delta \frac{d}{d\lambda}} x^i \Big|_P ,$$

$$x^i(A) = e^{\delta \frac{d}{d\mu}} e^{\delta \frac{d}{d\lambda}} x^i \Big|_P .$$

And for the route from P to B:

$$x^i(B) = e^{\delta \frac{d}{d\lambda}} e^{\delta \frac{d}{d\mu}} x^i \Big|_P .$$

The coordinate difference between points A and B is therefore:

$$x^i(B) - x^i(A) = e^{\delta \frac{d}{d\lambda}} e^{\delta \frac{d}{d\mu}} x^i \Big|_P - e^{\delta \frac{d}{d\mu}} e^{\delta \frac{d}{d\lambda}} x^i \Big|_P = \left[e^{\delta \frac{d}{d\lambda}}, e^{\delta \frac{d}{d\mu}} \right] x^i \Big|_P .$$

We develop the commutator into a Taylor series up to the second order and get:

$$\left[e^{\delta \frac{d}{d\lambda}}, e^{\delta \frac{d}{d\mu}} \right] = \left[1 + \delta \frac{d}{d\lambda} + \frac{1}{2!} \delta^2 \frac{d^2}{d\lambda^2} + \cdots, 1 + \delta \frac{d}{d\mu} + \frac{1}{2!} \delta^2 \frac{d^2}{d\mu^2} + \cdots \right]$$

$$= \delta^2 \left[\frac{d}{d\lambda}, \frac{d}{d\mu} \right] + \mathcal{O}(\delta^3).$$

And therefore with $\frac{dx^i}{d\lambda} = u^i(x^i)$, $\frac{dx^i}{d\mu} = v^i(x^i)$, (7.19), we get

$$\begin{aligned}
x^i(B) - x^i(A) &= \delta^2 \left[\frac{d}{d\lambda}, \frac{d}{d\mu} \right] x^i \Big|_P + \mathcal{O}(\delta^3) \\
&= \delta^2 \left[\vec{u}, \vec{v} \right]^i + \mathcal{O}(\delta^3).
\end{aligned} \tag{7.22}$$

almost a mathematical formulation of the Figure 7.7.

As mentioned above, equation (7.21), the commutator of two vector fields $\vec{u}$ and $\vec{v}$, the Lie bracket $[\vec{u}, \vec{v}]$, is again a *vector field*:

$$\begin{aligned}
[\vec{u}, \vec{v}] &= \vec{u}(\vec{v}) - \vec{v}(\vec{u}) \\
&= u^i \partial_i \left(v^j \partial_j \right) - v^j \partial_j \left(u^i \partial_i \right) \\
&= u^i \left(\partial_i v^j \right) \partial_j + u^i v^j \partial_i \partial_j - v^j \left(\partial_j u^i \right) \partial_i - v^j u^i \partial_j \partial_i \\
&= u^i \left(\partial_i v^j \right) \partial_j - \underbrace{v^j \left(\partial_j u^i \right) \partial_i}_{\text{dummy } i \leftrightarrow j} + \underbrace{u^i v^j \partial_i \partial_j - v^j u^i \partial_j \partial_i}_{=0} \\
&= \left(u^i \partial_i v^j - v^i \partial_i u^j \right) \partial_j, \\
[\vec{u}, \vec{v}]^j &= u^i \partial_i v^j - v^i \partial_i u^j .
\end{aligned} \tag{7.23}$$

The Lie bracket $[\vec{u}, \vec{v}]$ (7.23) is a tensor and transforms like a tensor, despite the obvious presence of partial derivatives (not tensorial, equation (6.1)). We now want to verify this:

$$[\vec{u}, \vec{v}]^{j'} = u^{i'} \partial_{i'} v^{j'} - v^{i'} \partial_{i'} u^{j'}$$

$$\left\{ u^{i'} \partial_{i'} = u^i \partial_i, \; v^{i'} \partial_{i'} = v^i \partial_i, \text{contraction, see equation (4.55)} \right\}$$

$$= u^i \partial_i \left(\frac{\partial x^{j'}}{\partial x^j} v^j \right) - v^i \partial_i \left(\frac{\partial x^{j'}}{\partial x^j} u^j \right)$$

$$= \left(u^i \left(\partial_i v^j \right) \frac{\partial x^{j'}}{\partial x^j} + u^i v^j \frac{\partial}{\partial x^i} \frac{\partial x^{j'}}{\partial x^j} \right) - \left(v^i \left(\partial_i u^j \right) \frac{\partial x^{j'}}{\partial x^j} + \underbrace{v^i u^j \frac{\partial}{\partial x^i} \frac{\partial x^{j'}}{\partial x^j}}_{\text{dummy } i \leftrightarrow j} \right)$$

$$= \left(u^i \left(\partial_i v^j \right) \frac{\partial x^{j'}}{\partial x^j} + u^i v^j \frac{\partial}{\partial x^i} \frac{\partial x^{j'}}{\partial x^j} \right) - \left(v^i \left(\partial_i u^j \right) \frac{\partial x^{j'}}{\partial x^j} + v^j u^i \frac{\partial}{\partial x^j} \frac{\partial x^{j'}}{\partial x^i} \right)$$

$$\left\{ \frac{\partial}{\partial x^j} \frac{\partial x^{j'}}{\partial x^i} = \frac{\partial}{\partial x^i} \frac{\partial x^{j'}}{\partial x^j}, \text{partial derivatives commute} \right\}$$

$$= u^i \left(\partial_i v^j \right) \frac{\partial x^{j'}}{\partial x^j} - v^i \left(\partial_i u^j \right) \frac{\partial x^{j'}}{\partial x^j}$$

$$= \frac{\partial x^{j'}}{\partial x^j} \left(u^i \partial_i v^j - v^i \partial_i u^j \right)$$

$$= \frac{\partial x^{j'}}{\partial x^j} [\vec{u}, \vec{v}]^j \, .$$

Remark 7.6
It is noteworthy that the Lie bracket $[\vec{u}, \vec{v}]$ is a *tensorial derivative operator* defined *without* any metric or connection.

■

In Section 7.2 we have shown that a vector or vector field acts like a derivative of a smooth function f on a manifold M, equation (7.4). Transferred to the Lie bracket this means

$$\vec{w} = [\vec{u}, \vec{v}], \quad \vec{w} f = \vec{u} \left(\vec{v} f \right) - \vec{v} \left(\vec{u} f \right). \tag{7.24}$$

The Lie bracket $\vec{w}$ satisfies the conditions of a vector space (see Box 7.2), in particular the Leibniz rule

$$\vec{w}(fg) = \left(\vec{w} f \right) g + f \left(\vec{w} g \right). \tag{7.25}$$

Box 7.7   *Lie derivative*

Let us examine the expression $[\vec{u}, f\vec{v}]$, where $\vec{u}$, $\vec{v}$ are vector fields and f is an arbitrary smooth function. Our starting point is equation (7.23).

$$
\begin{aligned}
[\vec{u}, f\vec{v}]^j &= u^i \partial_i \left(fv^j\right) - fv^i \partial_i u^j \\
&= u^i \left(\partial_i f\right) v^j + f u^i \partial_i v^j - f v^i \partial_i u^j \\
&= u^i \left(\partial_i f\right) v^j + f \left[\vec{u}, \vec{v}\right]^j .
\end{aligned}
$$

As a result one can see the Leibniz rule, if one reads it as a derivation with respect to $\vec{u}$:

$$
u^i \left(\partial_i f\right) \equiv \text{directional derivative of } f \text{ with respect to } \vec{u} = u^i \partial_i,
$$
$$
[\vec{u}, \vec{v}] \equiv \text{directional derivative of } \vec{v} \text{ with respect to } \vec{u}.
$$

The Lie derivative with respect to $\vec{u}$ is denoted $\mathcal{L}_{\vec{u}}(\ )$. And in this manner we write the above result as follows:

$$
\begin{aligned}
[\vec{u}, f\vec{v}] &= \vec{v}\,\mathcal{L}_{\vec{u}} f + f\,\mathcal{L}_{\vec{u}}\vec{v}, \\
\mathcal{L}_{\vec{u}} f &= u^i \left(\partial_i f\right) = \vec{u} \cdot \nabla f, \\
\mathcal{L}_{\vec{u}}\vec{v} &= [\vec{u}, \vec{v}] .
\end{aligned}
$$

7.12   Torsion Tensor

If the Lie bracket $[\vec{u}, \vec{v}] = \vec{u}(\vec{v}) - \vec{v}(\vec{u})$, the flow curves, and $(\nabla_{\vec{u}}\vec{v} - \nabla_{\vec{v}}\vec{u})$, the covariant transport, lead to an identical vector field,

$$
\nabla_{\vec{u}}\vec{v} - \nabla_{\vec{v}}\vec{u} = [\vec{u}, \vec{v}] , \tag{7.26}
$$

then one speaks of a *symmetric connection*. Both vector fields, $\nabla_{\vec{u}}\vec{v} - \nabla_{\vec{v}}\vec{u}$ and $[\vec{u}, \vec{v}]$, are antisymmetric in $\vec{u}$ and $\vec{v}$.

Box 7.8

With a *coordinate basis*, equation (7.26) states that the connection coefficients Γ^k_{ij} are *symmetric*. We use equation (6.8) for the LHS of (7.26) and for the RHS equation (7.23), which are based on a coordinate basis. And

---

then we obtain for the j-component of (7.26):

$$(\nabla_{\vec{u}}\vec{v} - \nabla_{\vec{v}}\vec{u})^j = [\vec{u},\vec{v}]^j\,,$$

$$u^i\left(\partial_i v^j + \Gamma^j_{ik}v^k\right) - v^i\left(\partial_i u^j + \Gamma^j_{ik}u^k\right) = u^i\partial_i v^j - v^i\partial_i u^j,$$

$$u^i\Gamma^j_{ik}v^k - \underbrace{v^i\Gamma^j_{ik}u^k}_{i\leftrightarrow k} = 0,$$

$$u^i\Gamma^j_{ik}v^k - v^k\Gamma^j_{ki}u^i = 0,$$

$$u^i v^k\left(\Gamma^j_{ik} - \Gamma^j_{ki}\right) = 0,$$

$$\Longrightarrow$$

$$\underbrace{\Gamma^j_{ik} = \Gamma^j_{ki}}_{j\leftrightarrow k},\ \text{or }\ \Gamma^k_{ij} = \Gamma^k_{ji}.$$

A symmetrical connection is also called torsion-free. If a connection is
non-symmetrical, then based on equation (7.26), the *torsion tensor* is defined
as follows (the components of a tensor are determined by the action of the
corresponding basis vectors, see Section 3.2, (3.10)) :

$$\mathbf{T}(\ ;\ ) = T^k{}_{ij}\,\vec{e}_i \otimes \vec{e}_j \otimes \tilde{e}^k,$$
$$T^k{}_{ij} = \left(\nabla_{\vec{e}_i}\vec{e}_j - \nabla_{\vec{e}_j}\vec{e}_i - [\vec{e}_i,\vec{e}_j]\right)^k = \Gamma^k{}_{ij} - \Gamma^k{}_{ji}. \tag{7.27}$$

The torsion tensor $\mathbf{T}$ is a $(1,2)$-tensor, two slots for vector fields to be tested for
connection symmetry and one slot for a one-form to determine the component
of the outcome vector field.

7.13   The commutator of two covariant derivatives

The covariant transport (derivative) $\nabla_{\partial_i}v^k$ of a vector $\vec{v}$ in the direction of
coordinate x^i is the difference to the parallel-transport of the vector $\vec{v}$, since
$\nabla_{\partial_i}v^k = 0$. Are two covariant transports (derivations), $\nabla_{\partial_j}\nabla_{\partial_i}$, carried out
one after the other symmetrical, as the ordinary partial derivations, $\partial_j\partial_i$, are?

We can verify that by analysing the commutator $\left[\nabla_{\partial_j}, \nabla_{\partial_i}\right] v^k =$

$$
\begin{aligned}
&= \nabla_{\partial_j}\nabla_{\partial_i}v^k - \nabla_{\partial_i}\nabla_{\partial_j}v^k \\
&= \nabla_{\partial_j} \underbrace{\left(\nabla_{\partial_i}v^k\right)}_{\text{transported first}} - (i \leftrightarrow j) \\
&= \partial_j\left(\nabla_{\partial_i}v^k\right) + \Gamma^k_{js}\left(\nabla_{\partial_i}v^s\right) - \Gamma^l_{ji}\left(\nabla_{\partial_l}v^k\right) - (i \leftrightarrow j) \\
&= \partial_j\left(\partial_i v^k + \Gamma^k_{is}v^s\right) + \Gamma^k_{js}\left(\partial_i v^s + \Gamma^s_{il}v^l\right) - \Gamma^l_{ji}\left(\nabla_{\partial_l}v^k\right) - (i \leftrightarrow j) \\
&= \partial_j\partial_i v^k + \left(\partial_j\Gamma^k_{is}\right)v^s + \Gamma^k_{is}\left(\partial_j v^s\right) + \Gamma^k_{js}\left(\partial_i v^s\right) + \Gamma^k_{js}\Gamma^s_{il}v^l - \Gamma^l_{ji}\left(\nabla_{\partial_l}v^k\right) \\
&\quad - (i \leftrightarrow j) \\
&= \left(\partial_j\partial_i v^k + \left(\partial_j\Gamma^k_{is}\right)v^s + \Gamma^k_{is}\left(\partial_j v^s\right) + \Gamma^k_{js}\left(\partial_i v^s\right) + \underbrace{\Gamma^k_{js}\Gamma^s_{il}v^l}_{s \leftrightarrow l} - \Gamma^l_{ji}\left(\nabla_{\partial_l}v^k\right)\right) - \\
&\quad - \left(\partial_i\partial_j v^k + \left(\partial_i\Gamma^k_{js}\right)v^s + \Gamma^k_{js}\left(\partial_i v^s\right) + \Gamma^k_{is}\left(\partial_j v^s\right) + \underbrace{\Gamma^k_{is}\Gamma^s_{jl}v^l}_{s \leftrightarrow l} - \Gamma^l_{ij}\left(\nabla_{\partial_l}v^k\right)\right) \\
&= \underbrace{\left(\partial_j\Gamma^k_{is} - \partial_i\Gamma^k_{js} + \Gamma^k_{jl}\Gamma^l_{is} - \Gamma^k_{il}\Gamma^l_{js}\right)}_{\text{Riemann Tensor } R^k{}_{sji}} v^s - \underbrace{\left(\Gamma^l_{ji} - \Gamma^l_{ij}\right)}_{\text{Torsion tensor } T^l_{ji}} \nabla_{\partial_l}v^k.
\end{aligned}
$$

And therefore

$$
\left[\nabla_{\partial_j}, \nabla_{\partial_i}\right] v^k = R^k{}_{sji}v^s - T^l_{ji}\nabla_{\partial_l}v^k. \tag{7.28}
$$

In the case of torsional freedom (coordinates basis, Section 6.4), the commutator of the covariant derivative of a vector field is only a function of the Riemann tensor, which we already got to know in Section 7.7 regarding a parallel transport of a vector field via the difference of two paths (commutator).

$$
\left[\nabla_{\partial_j}, \nabla_{\partial_i}\right] v^k = R^k{}_{sji}v^s \tag{7.29}
$$

The Riemann tensor is antisymmetric on index places 3 and 4 (j and i), Section 7.8, and this is consistent with $\left[\nabla_{\partial_j}, \nabla_{\partial_i}\right] = -\left[\nabla_{\partial_i}, \nabla_{\partial_j}\right]$.

7.14 Geodesic deviation

Two non-parallel straight lines in flat space $\mathfrak{g}_1(\rho)$ and $\mathfrak{g}_2(\rho)$, parametrized with ρ, change their distance to each other as one moves along the straight line, see Figure 7.8. This separation vector $\vec{s}$ is a linear function of the parameter ρ.

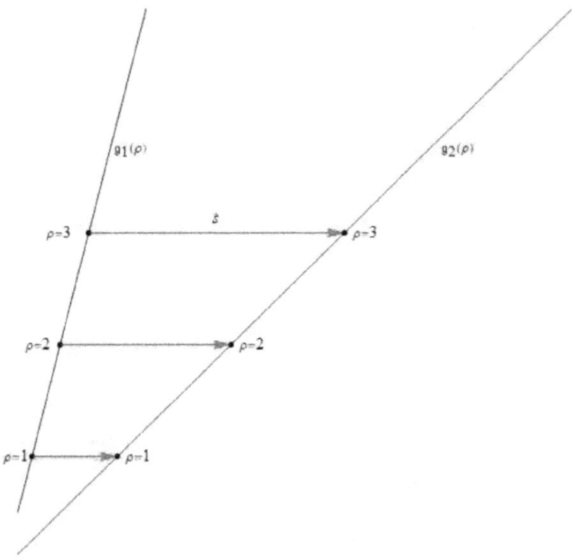

Figure 7.8: Two straight lines $\mathfrak{g}_1(\rho)$ and $\mathfrak{g}_2(\rho)$ in Euclidean space with their separation vector $\vec{s}$.

His "velocity" $\frac{d\vec{s}}{d\rho}$ is constant. And therefore has no "acceleration": $\frac{d^2\vec{s}}{d\rho^2} = 0$. In curved space, however, the distance vector of two geodesics does have an "acceleration", since the distance vector does not change linearly to the movement along the geodesics.

We now consider[1] a region of a manifold M with a set of *initially parallel* geodesics $\{\mathfrak{g}_\sigma(\lambda)\}$, $\lambda, \sigma \in \mathbb{R}$, with a separation parameter σ, Figure 7.9. The coordinates of this two-dimensional surface of M are $x^i(\lambda, \sigma)$. The coordinate lines $x^i(\lambda = \text{constant}, \sigma)$ connect points on the geodesics with equal λ-values. The tangential vector of the geodesic coordinates $u^k = \frac{\partial x^k}{\partial \lambda}$ and the tangential vector of the connection coordinates $s^k = \frac{\partial x^k}{\partial \sigma}$ form a coordinate basis and

---

[1] Argumentation according to Sean M. Carroll, *Spacetime and Geometry*, 2016, [Car19]

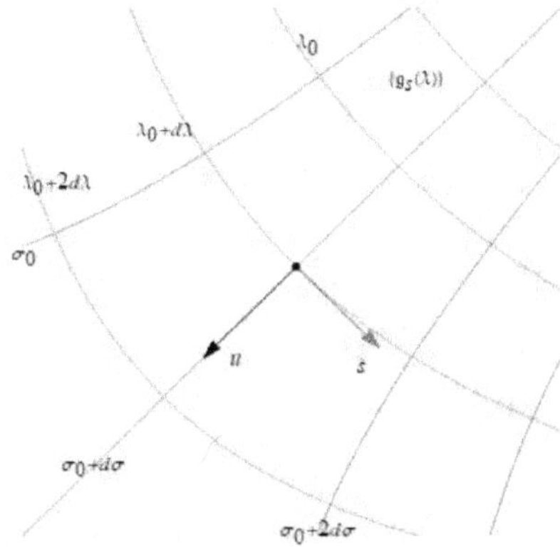

Figure 7.9: A set of initially parallel geodesics with tangent vector $\vec{u}$ and separation vector $\vec{s}$.

therefore applies:

$$[\vec{u}, \vec{s}] = \vec{0}. \tag{7.30}$$

We consider a symmetric connection, then equation (7.26) applies:

$$\nabla_{\vec{u}}\vec{s} = \nabla_{\vec{s}}\vec{u},$$
$$u^i \nabla_i s^k = s^i \nabla_i u^k. \tag{7.31}$$

We define with,

$$v^k = (\nabla_{\vec{u}}\vec{s})^k = u^i \nabla_i s^k, \tag{7.32}$$

the *relative separation velocity* between the geodesics, and with,

$$a^k = (\nabla_{\vec{u}}\vec{v})^k = u^i \nabla_i v^k = u^i \nabla_i \left(u^j \nabla_j s^k \right), \tag{7.33}$$

the *relative separation acceleration* of the geodesics.

In the following calculation of a^k a double differential $\nabla\nabla$. We treat this with the 'reverse' Leibniz rule. For this see Box 7.9.

Box 7.9

Leibniz rule (LR):

$$Ad_x\left(Bd_y\right) = A\left(d_x B\right) d_y + AB d_x d_y.$$

reverse Leibniz rule (rLR):

$$AB d_x d_y = Ad_x\left(Bd_y\right) - A\left(d_x B\right) d_y.$$

Starting from equation (7.33) we will derive an expression for the relative separation acceleration a^k of the geodesics:

$$a^k = u^i \nabla_i\left(u^j \nabla_j s^k\right) \overset{(7.31)}{=} u^i \nabla_i\left(s^j \nabla_j u^k\right)$$

$$\overset{\text{LR}}{=} \left(u^i \nabla_i s^j\right) \nabla_j u^k + u^i s^j \left(\nabla_i \nabla_j u^k\right)$$

$$\overset{(7.29)}{=} \left(u^i \nabla_i s^j\right) \nabla_j u^k + u^i s^j \left(\nabla_j \nabla_i u^k + R^k_{\;sij} u^s\right)$$

$$= \left(u^i \nabla_i s^j\right) \nabla_j u^k + \left[\underbrace{u^i s^j}_{=s^j u^i} \left(\nabla_j \nabla_i u^k\right)\right] + R^k_{\;sij} u^i s^j u^s$$

$$\overset{\text{rLR}}{=} \left(u^i \nabla_i s^j\right) \nabla_j u^k + \left[s^j \nabla_j \underbrace{\left(u^i \nabla_i u^k\right)}_{=0,(7.12)} - \left(s^j \nabla_j u^i\right) \nabla_i u^k\right] + R^k_{\;sij} u^s u^i s^j$$

$$= \left(u^i \nabla_i s^j\right) \nabla_j u^k - \underbrace{\left(s^j \nabla_j u^i\right) \nabla_i u^k}_{i \leftrightarrow j} + R^k_{\;sij} u^s u^i s^j$$

$$= \underbrace{\left(u^i \nabla_i s^j - s^i \nabla_i u^j\right)}_{=0,(7.31)} \nabla_j u^k + R^k_{\;sij} u^s u^i s^j$$

$$= R^k_{\;sij} u^s u^i s^j.$$

The result is the so-called *geodesic deviation equation*:

$$a^k = \frac{d^2 s^k}{d\lambda^2} = R^k_{\;sij} u^s u^i s^j. \tag{7.34}$$

Example 7.6

We want to study the geodesic deviation equation (7.34) for a 2-shere $\{\theta, \phi\}$. We will assume geodesics along the θ-coordinate line, i.e. $u^\phi = 0$. Since the tangential vector $\vec{u}$ of a geodesic is constant in magnitude and direction (see definition Section 7.6), the following applies: $u^\theta =$ constant.

Using the components of the Riemann tensor for a 2-sphere (see Summary 7.1) we obtain two differential equations:

$$
\begin{aligned}
\frac{d^2 s^\theta}{d\lambda^2} &= R^\theta{}_{sij} u^s u^i s^j \\
&= R^\theta{}_{\phi\phi\theta} u^\phi u^\phi s^\theta + R^\theta{}_{\phi\theta\phi} u^\phi u^\theta s^\phi \\
&\overset{u^\phi=0}{=} 0,
\end{aligned}
$$

and second

$$
\begin{aligned}
\frac{d^2 s^\phi}{d\lambda^2} &= R^\phi{}_{sij} u^s u^i s^j \\
&= R^\phi{}_{\theta\phi\theta} u^\theta \underbrace{u^\phi}_{=0} s^\theta + \underbrace{R^\phi{}_{\theta\theta\phi}}_{=-1} u^\theta u^\theta s^\phi \\
&= -\left(u^\theta\right)^2 s^\phi.
\end{aligned}
$$

The first equation states that the separation vector $\vec{s}$, which connects equal λ-values of the geodesics, moves with constant speed along the θ-coordinate. The second equation has a trigonometric solution:

$$
s^\phi = s_0^\phi \cos\left(u^\theta \lambda\right).
$$

The interpretation of the result is that the separation coordinate s^ϕ between two geodesics performs an oscillation when the geodesic, parametrised by λ, extends from the equator to the pole and beyond to the equator again and so on.

■

7.15 Bianchi identies

We already know a Bianchi identity, Section 7.8, (7.16),

$$R_{ksji} + R_{kjis} + R_{kisj} = 0,$$

which results from the cyclic permutation of the last three indices of the low-index representation of the Riemann tensor R_{ksji}, (7.15). A second Bianchi identity is known, which is obtained from the differentiation of the low-index representation of the Riemann tensor R_{ksji}, (7.15), with respect to a coordinate x^l.

$$R_{ksji} = \frac{1}{2}\left(\partial_j \partial_s g_{ki} - \partial_j \partial_k g_{is} - \partial_i \partial_s g_{kj} + \partial_i \partial_k g_{js}\right),$$

$$\partial_l R_{ksji} = \frac{1}{2}\left(\partial_l \partial_j \partial_s g_{ki} - \partial_l \partial_j \partial_k g_{is} - \partial_l \partial_i \partial_s g_{kj} + \partial_l \partial_i \partial_k g_{js}\right).$$

Shifting the indices 1,4 and 5 cyclically to the right with subsequent addition, we obtain the following result

$$\partial_l R_{ksji} + \partial_i R_{kslj} + \partial_j R_{ksil} = 0.$$

In inertial frame also applies:

$$\nabla_l R_{ksji} + \nabla_i R_{kslj} + \nabla_j R_{ksil} = 0, \tag{7.35}$$

Since a tensor retains its validity in any coordinate system, the above expression, the so-called *Bianchi identity* 2, is valid in any frame. Because of symmetry of the $(12, 34)$ index swap, (7.16), we get a cyclical rotation on the first three indices:

$$\nabla_l R_{ksji} + \nabla_i R_{kslj} + \nabla_j R_{ksil} = \nabla_{[l} R_{ji]ks} = 0. \tag{7.36}$$

7.16 Ricci tensor

An evaluation of the contraction properties of the Riemann tensor $R^k_{\ sji}$ due to its symmetry properties of the indices, equation (7.16), yields the following result.

Remark 7.7

A contraction can only be performed via an upper index and a lower index.
Never between two upper indices or two lower indices!

∎

$\text{contraction}(1,2) : R^k{}_{kji} = 0,$

$\text{contraction}(1,3) : R^k{}_{ski},$

$\text{contraction}(1,4) : R^k{}_{sjk} = -R^k{}_{skj} \equiv -\text{contraction}(1,3),$

$\text{contraction}(2,3) : R_k{}^s{}_{si} = g^{as} R_{kasi} = -g^{as} R_{aksi} = -R^s{}_{ksi} \equiv -\text{contraction}(1,3),$

$\text{contraction}(2,4) : R_k{}^s{}_{js} = g^{as} R_{kajs} = -g^{as} R_{akjs}$

$$= -g^{as}(-R_{aksj}) = R^s{}_{ksj} \equiv \text{contraction}(1,3),$$

$\text{contraction}(3,4) : R_{ks}{}^j{}_j = 0.$

This leaves only the $(1,3)$ contraction for the Riemann tensor as the relevant contraction with the result $\pm R_{si}$. Because of $R^k{}_{ski} = R_{ki}{}^k{}_s$, Index $12-34$ swap, the remaining tensor is symmetrical and is called *Ricci tensor*:

$$R^k{}_{ski} = R_{si} = R_{is}. \tag{7.37}$$

The *Ricci scalar* R, a scale for the curvature of space, is defined as

$$R = g^{si} R_{si}. \tag{7.38}$$

Example 7.7 Ricci tensor R_{si} and Ricci scalar R for sphere $\{\theta, \phi\}$

With the results from the Summary 7.1 and equation (7.37) we obtain by means of a $(1,3)$ contraction over *all* indices the Ricci tensor.

$$R_{\phi\phi} = R^\theta{}_{\phi\theta\phi} + R^\phi{}_{\phi\phi\phi} = \sin^2\theta + 0 = \sin^2\theta,$$

$$R_{\theta\theta} = R^\theta{}_{\theta\theta\theta} + R^\phi{}_{\theta\phi\theta} = 0 + 1 = 1,$$

$$R_{\theta\phi} = R^\theta{}_{\theta\theta\phi} + R^\phi{}_{\theta\phi\phi} = 0 + 0 = 0,$$

$$R_{\phi\theta} = R_{\theta\phi} = 0.$$

The Ricci tensor R_{si} for a sphere shown as matrix:

$$R_{si} = \begin{pmatrix} 1 & 0 \\ 0 & \sin^2\theta \end{pmatrix}.$$

The trace of this matrix is the Ricci scalar of a sphere. But be careful, this is not $1 + \sin^2\theta$, because a trace is always a contraction (see Section 4.11). Therefore we have to raise one index to be able to perform the contraction, the summation over the diagonal.

$$R = R_i^i = g^{si}R_{si} = g^{\theta\theta}R_{\theta\theta} + g^{\phi\phi}R_{\phi\phi}$$
$$= \frac{1}{r^2}\cdot 1 + \frac{1}{r^2\sin^2\theta}\cdot\sin^2\theta = \frac{2}{r^2}.$$

$\blacksquare$

Chapter 8

Differential forms

8.1 Antisymmetric tensors

Let us start with an example. A $(0,2)$-tensor $\mathbf{T}$ has the property $\mathbf{T}(\vec{u}, \vec{u}) = 0$, for any vector $\vec{u}$. The investigation of this feature of a tensor results in an antisymmetry if the arguments are interchanged. Without restriction of generality we assume that $\vec{u} = \vec{v} + \vec{w}$, and then we get:

$$\mathbf{T}(\vec{u}, \vec{u}) = \mathbf{T}(\vec{v} + \vec{w}, \vec{v} + \vec{w}) = \mathbf{T}(\vec{v}, \vec{v}) + \mathbf{T}(\vec{v}, \vec{w}) + \mathbf{T}(\vec{w}, \vec{v}) + \mathbf{T}(\vec{w}, \vec{w})$$
$$= 0 + \mathbf{T}(\vec{v}, \vec{w}) + \mathbf{T}(\vec{w}, \vec{v}) + 0 = 0,$$
$$\mathbf{T}(\vec{v}, \vec{w}) = -\mathbf{T}(\vec{w}, \vec{v}).$$

And hence the following definition applies: A $(0,2)$-tensor is *antisymmetric* if its value changes sign when the arguments are interchanged:

$$\mathbf{T}(\vec{u}, \vec{u}) = 0,$$
$$\mathbf{T}(\vec{u}, \vec{v}) = -\mathbf{T}(\vec{v}, \vec{u}),$$
$$\mathbf{T}(\vec{e}_i, \vec{e}_j) = -\mathbf{T}(\vec{e}_j, \vec{e}_i),$$
$$T_{ij} = -T_{ji},$$
$$T_{ii} = 0.$$

For $(0,p)$-tensors, $p \geq 3$. it must be indicated on which indices there is an antisymmetry. Therefore the symbol of a square bracket has been introduced:

$$T_{i[jk]lm\ldots}, \text{ antisymmetric with respect to the second and third index.}$$

A tensor is called *completely antisymmetric* if it changes the sign when any two indices or arguments are exchanged. Its representation is (see Section 4.11, (4.51)):

$$T_{[ijk]}, \ T_{[ij\cdots n]}.$$

Box 8.1

The components of a completely antisymmetric $(0, p)$-tensor $T_{i_1 i_2 \cdots i_p}$ have the following characteristics:
1) They do not change their value when the indices are shift cyclically.
2) When two indices are interchanged, they change their sign.
Using the example of an $(0, 3)$-tensor:

$$T_{ijk} = T_{jki} = T_{kij} = -T_{jik} = -T_{kji} = -T_{ikj}.$$

Example 8.1

A is an antisymmetric $(0, 2)$-tensor, $A_{ij} = -A_{ji}$, and **B** is a symmetric $(2, 0)$-tensor, $B^{ij} = B^{ji}$. A contraction of these two tensors leads to the following result:

$$A_{ij}B^{ij} = -A_{ji}B^{ji} \xrightarrow[i \leftrightarrow j]{\text{dummy index}} = -A_{ij}B^{ij},$$

$$A_{ij}B^{ij} = -A_{ij}B^{ij} \implies A_{ij}B^{ij} = 0.$$

This indicates that antisymmetric indices select out the symmetric parts of the contraction counterpart during *contraction* and what remains are the antisymmetric parts of the contraction counterpart:

$$A_{[ij]}B^{ij} = A_{[ij]}B^{[ij]}. \tag{8.1}$$

If an anti-symmetric $(0, 2)$-tensor A_{ij} consists of summands, then the *sum* does not contain any symmetrical tensors. This is shown by an example with two summands B_{ij} and C_{ij}. Let B be antisymmetric and C be symmetric. Because of equation (4.51) applies.

$$
\begin{aligned}
A_{[ij]} &= \frac{1}{2}\left(A_{ij} - A_{ji}\right) \\
&= \frac{1}{2}\left((B_{ij} + C_{ij}) - (B_{ji} + C_{ji})\right) \\
&= \frac{1}{2}\left((B_{ij} - B_{ji}) - \underbrace{(C_{ij} - C_{ji})}_{=0,\, C_{ij}=C_{ji}}\right) \\
&= B_{[ij]}.
\end{aligned}
\tag{8.2}
$$

$\blacksquare$

The maximum number of *free components* of a completely antisymmetric $(0,p)$-tensor, $p \leq n$, in an n-dimensional manifold M is at most :

$$
C_p^n = \frac{n!}{p!(n-p)!} = \binom{n}{p}.
\tag{8.3}
$$

Consider: You choose p different numbers from a set $(1, \dots, n)$. In the first step, $p = 1$, you have n possibilities. In the second step, $p = 2$, you have only $(n-1)$ possibilities, because the indices must be different for completely antisymmetric tensors. After p selection steps one has selected $n(n-1)(n-2)\cdots(n-p+1) = n!/(n-p)!$ possibilities. The order of the selection is irrelevant, since only the sign changes when two indices are swapped. So $n!/(n-p)!$ must still be divided by $p!$. C_p^n is therefore the binomial coefficient.

8.2 Differential forms

In Section 2.2 we have introduced the 1-form,

$$
\tilde{p} = p_i \tilde{e}^i \in \text{dual space } \mathcal{V}^\star.
$$

A 1-form maps a vector as argument to a real number:

$$
\tilde{p}(\vec{v}) = p_i \tilde{e}^i(\vec{v}) = p_i v^i \in \mathbb{R}.
$$

We are introducing a new symbol for the basis of dual space:

$$
\tilde{e}^i \equiv d\tilde{x}^i, \quad \text{one-form} : \quad \tilde{p} = p_i d\tilde{x}^i.
\tag{8.4}
$$

This is also the origin of the term "differential forms" for the dual-vector forms in this chapter. The usefulness of this will be demonstrated in the next few sections.

Definition 8.1

A p-form is a *completely antisymmetric* $(0, p)$-tensor ($p \geq 2$). The p-form vector space is represented by the symbol $\wedge^p \mathcal{V}^\star$. The number p is the degree of the form. The components $p_{i_1 i_2 \cdots i_p}$ of a p-form, are determined by the action on the corresponding basis vectors $\{\vec{e}_{i_1}, \vec{e}_{i_2}, \ldots, \vec{e}_{i_p}\} \in \mathcal{V}$, see Section 3.2, (3.10).

A 0-form is a scalar, a function. A 1-form is the dual vector $\tilde{p}$ known from Chapter 2.

∎

In contrast to the so far mentioned elements of the vector space $\mathcal{V}^\star$, namely one-forms (covectors), it is possible to form arbitrary *products* from p-forms. The so-called wedge or exterior product. Symbolised by $\wedge$. $\wedge^p \mathcal{V}^\star$ is an *exterior* algebra on the vector space $\mathcal{V}^\star$. The set of all forms with the rules of antisymmetric multiplication $\wedge$ is called a *Grassmann algebra* or *exterior algebra*. It is defined via the dual basis vectors $d\tilde{x}^i$.

Definition 8.2

The wedge product (exterior product), $\wedge$, is defined as *anti-symmetrized tensor product* of the dual basis $d\tilde{x}^i$:

$$d\tilde{x}^i \wedge d\tilde{x}^j = d\tilde{x}^i \otimes d\tilde{x}^j - d\tilde{x}^j \otimes d\tilde{x}^i, \tag{8.5a}$$

$$d\tilde{x}^i \wedge d\tilde{x}^j = -d\tilde{x}^j \wedge d\tilde{x}^i. \tag{8.5b}$$

A wedge product with a 0-form f is defined by

$$f \wedge d\tilde{x}^i = f \, d\tilde{x}^i.$$

∎

As a brief example, let us form the (wedge)$\wedge$-product of two 1-forms $\tilde{p}$ and $\tilde{q}$ in a 3-dimensional vector space, using (8.5b). In the first step, the indices run through their possible values according to the contraction rules, and only in the second step is use made of the antisymmetry of the wedge product,

(8.5b). To make the calculation clearer, the wedge symbol $\wedge$ is omitted from the second line onwards.

$$
\begin{aligned}
\tilde{p} \wedge \tilde{q} = \left(p_i d\tilde{x}^i\right) \wedge \left(q_j d\tilde{x}^j\right) &= p_i q_j d\tilde{x}^i \wedge d\tilde{x}^j \\
&= p_1 q_2 d\tilde{x}^1 d\tilde{x}^2 + p_2 q_1 d\tilde{x}^2 d\tilde{x}^1 + \\
&\quad + p_1 q_3 d\tilde{x}^1 d\tilde{x}^3 + p_3 q_1 d\tilde{x}^3 d\tilde{x}^1 + \\
&\quad + p_2 q_3 d\tilde{x}^2 d\tilde{x}^3 + p_3 q_2 d\tilde{x}^3 d\tilde{x}^2 \\
&= p_1 q_2 d\tilde{x}^1 d\tilde{x}^2 - p_2 q_1 d\tilde{x}^1 d\tilde{x}^2 + \\
&\quad + p_1 q_3 d\tilde{x}^1 d\tilde{x}^3 - p_3 q_1 d\tilde{x}^1 d\tilde{x}^3 + \\
&\quad + p_2 q_3 d\tilde{x}^2 d\tilde{x}^3 - p_3 q_2 d\tilde{x}^3 d\tilde{x}^2 \\
&= \left(p_1 q_2 - p_2 q_1\right) d\tilde{x}^1 d\tilde{x}^2 + \left(p_1 q_3 - p_3 q_1\right) d\tilde{x}^1 d\tilde{x}^3 + \left(p_2 q_3 - p_3 q_2\right) d\tilde{x}^2 d\tilde{x}^3 .
\end{aligned}
$$
$$(8.6)$$

The result is a 2-form $\tilde{r} = \tilde{p} \wedge \tilde{q} \in \wedge^2 \mathcal{V}^\star$. You can see the anti-symmetrical components $r_{ij} = p_i q_j - p_j q_i$ with the corresponding basis $d\tilde{x}^i \wedge d\tilde{x}^j \equiv d\tilde{x}^i d\tilde{x}^j$. Thus it is also shown that a 2-form in a 3-dimensional vector space has 3 free components, as expected from formula (8.3) , $\binom{3}{2} = 3$.

Usually and in the following, the basis

$$
\left\{d\tilde{x}^I\right\} = \left\{d\tilde{x}^{i_1} \wedge \ldots \wedge d\tilde{x}^{i_p}\right\}, \; i_1 < \ldots < i_p, \; p \leq n, \tag{8.7}
$$

is chosen for a p-form of the vector space $\wedge^p \mathcal{V}^\star$ on an n-dimensional manifold M with the index set $I = \{i_1, \ldots, i_p\}$.

In summary, the following rules apply based on definitions 8.1 and 8.2. Let $\tilde{p}$ be a p-form, $\tilde{q}$ and $\tilde{r}$ be q-forms, $a, b, c \in \mathbb{R}$.

$$
\begin{aligned}
\tilde{p} \wedge \left(b\tilde{q} + c\tilde{r}\right) &= b\tilde{p} \wedge \tilde{q} + c\tilde{p} \wedge \tilde{r}, \\
\left(a\tilde{p} + b\tilde{q}\right) \wedge \tilde{r} &= a\tilde{p} \wedge \tilde{r} + b\tilde{q} \wedge \tilde{r}, \\
\left(\tilde{p} \wedge \tilde{q}\right) \wedge \tilde{r} &= \tilde{p} \wedge \left(\tilde{q} \wedge \tilde{r}\right), \\
\tilde{p} \wedge \tilde{q} &= (-1)^{pq} \tilde{q} \wedge \tilde{p}.
\end{aligned}
\tag{8.8}
$$

Box 8.2

On a manifold M ($n = 3$) there are *four* p-form vector spaces, namely $\wedge^0 \mathcal{V}^\star$, $\wedge^1 \mathcal{V}^\star$ $\wedge^2 \mathcal{V}^\star$ and $\wedge^3 \mathcal{V}^\star$.

$p = 0$: smooth function f, the basis is $f = 1$.

$p = 1$: $\tilde{a} = a_I d\tilde{x}^I = a_1 d\tilde{x}^1 + a_2 d\tilde{x}^2 + a_3 d\tilde{x}^3$.

$p = 2$: $\tilde{b} = b_I d\tilde{x}^I = b_{12} d\tilde{x}^1 \wedge d\tilde{x}^2 + b_{13} d\tilde{x}^1 \wedge d\tilde{x}^3 + b_{23} d\tilde{x}^2 \wedge d\tilde{x}^3$.

$p = 3$: $\tilde{c} = c_I d\tilde{x}^I = c_{123} \, d\tilde{x}^1 \wedge d\tilde{x}^2 \wedge d\tilde{x}^3$.

c_{ijk} are the components of a completely antisymmetric tensor, and with $n = 3$ there is only *one free component*, (8.3).

This clearly shows the possible *free components* of each p-form. 0-form: one, 1-form three, 2-form three and 3-form one. In total, $\sum_{p=0}^{p=3} C_p^3 = \sum_{p=0}^{p=3} \binom{3}{p} = 8$, (8.3).

A general formulation of a 2-form is

$$\tilde{b} = \frac{1}{2!} b_{[ij]} d\tilde{x}^i \wedge d\tilde{x}^j, \tag{8.9}$$

where the factor $\frac{1}{2!}$ is added to avoid double counting of the indices because of the antisymmetry of the tensor components. A restriction on the dimensionality of the vector space is not specified herein.

Example 8.2

We want to apply the 2-form of equation (8.9) to any two vectors. Use the *tensor product formulation* of the wedge product of the dual basis, equation (8.5a).

$$\tilde{b}\,(\vec{u}, \vec{v}) = \frac{1}{2!} b_{[ij]} d\tilde{x}^i \wedge d\tilde{x}^j \,(\vec{u}, \vec{v}) = \frac{1}{2!} b_{[ij]} \left(d\tilde{x}^i(\vec{u}) \otimes d\tilde{x}^j(\vec{v}) - d\tilde{x}^j(\vec{u}) \otimes d\tilde{x}^i(\vec{v}) \right)$$

$$= \frac{1}{2!} b_{[ij]} \left(u^i v^j - u^j v^i \right) = \frac{1}{2!} \left(b_{[ij]} u^i v^j \underbrace{- b_{[ij]}}_{b_{[ji]}} u^j v^i \right)$$

$$= \frac{1}{2!} \left(b_{[ij]} u^i v^j + \underbrace{b_{[ji]} u^j v^i}_{\text{dummy } i \leftrightarrow j} \right) = \frac{1}{2!} \left(b_{[ij]} u^i v^j + b_{[ij]} u^i v^j \right)$$

$$= b_{[ij]} u^i v^j. \qquad \blacksquare$$

If a p-form is fed with one vector only, it becomes a $(p-1)$-form. We want to show this with the 2-form from Example 8.2:

$$\tilde{b}\,(\vec{u},\ ) = \frac{1}{2!}b_{[ij]}d\tilde{x}^i \wedge d\tilde{x}^j\,(\vec{u},\ ) = \frac{1}{2!}b_{[ij]}\left(d\tilde{x}^i(\vec{u}) \otimes d\tilde{x}^j - d\tilde{x}^j(\vec{u}) \otimes d\tilde{x}^i\right)$$

$$= \frac{1}{2!}b_{[ij]}\left(u^i d\tilde{x}^j - u^j d\tilde{x}^i\right) = \frac{1}{2!}\left(b_{[ij]}u^i d\tilde{x}^j \underbrace{-b_{[ij]}}_{b_{[ji]}}u^j d\tilde{x}^i\right)$$

$$= \frac{1}{2!}\left(b_{[ij]}u^i d\tilde{x}^j + \underbrace{b_{[ji]}u^j d\tilde{x}^i}_{\text{dummy } i\leftrightarrow j}\right) = \frac{1}{2!}\left(b_{[ij]}u^i d\tilde{x}^j + b_{[ij]}u^i d\tilde{x}^j\right)$$

$$= b_{[ij]}u^i d\tilde{x}^j.$$

The result is obviously a 1-form. The index i is contracted $\rightarrow \mathbb{R}$. Although the vector $\vec{u}$ is only in the first slot of $d\tilde{x}^i \wedge d\tilde{x}^j\,(\vec{u},\ )$, it is contracted with each index of $b_{[ij]}$ due to the permutation of the $\wedge$-operation.

Box 8.3

The processing of three vectors by a 3-form is exemplified here. The basic representation of any 3-form is

$$\tilde{\gamma} = \frac{1}{3!}\gamma_{[ijk]}\,d\tilde{x}^i \wedge d\tilde{x}^j \wedge d\tilde{x}^k,$$

where the factor $\frac{1}{3!}$ is added to avoid double counting of the indices because of the *completely* antisymmetry of the tensor components.
Let's process $\tilde{\gamma}$ three vectors

$$\tilde{\gamma}\,(\vec{u},\vec{v},\vec{w}) = \frac{1}{3!}\gamma_{[ijk]}\,d\tilde{x}^i \wedge d\tilde{x}^j \wedge d\tilde{x}^k\,(\vec{u},\vec{v},\vec{w})$$

$$= \frac{1}{3!}\gamma_{[ijk]}(d\tilde{x}^i(\vec{u})d\tilde{x}^j(\vec{v})d\tilde{x}^k(\vec{w}) + d\tilde{x}^j(\vec{u})d\tilde{x}^k(\vec{v})d\tilde{x}^i(\vec{w}) +$$

$$+ d\tilde{x}^k(\vec{u})d\tilde{x}^i(\vec{v})d\tilde{x}^j(\vec{w}) - d\tilde{x}^j(\vec{u})d\tilde{x}^i(\vec{v})d\tilde{x}^k(\vec{w}) -$$

$$- d\tilde{x}^k(\vec{u})d\tilde{x}^j(\vec{v})d\tilde{x}^i(\vec{w}) - d\tilde{x}^i(\vec{u})d\tilde{x}^k(\vec{v})d\tilde{x}^j(\vec{w}))$$

$$= \frac{1}{3!}\gamma_{[ijk]}(u^i v^j w^k + u^j v^k w^i + u^k v^i w^j - u^j v^i w^k - u^k v^j w^i - u^i v^k w^j).$$

In order to make the calculation clearer, we consider each individual multiplication term separately, convert it in a way that all summands can be added. For a better overview, we omit the square brackets (antisymmetry of the indices) in the intermediate calculation.

$$\text{First term} : \gamma_{ijk}u^i v^j w^k,$$

$$\text{Second term} : \underbrace{\gamma_{ijk}}_{=\gamma_{jki}} u^j v^k w^i = \gamma_{jki}u^j v^k w^i = \underbrace{\gamma_{jki}u^j v^k w^i}_{j\to i, k\to j, i\to k} = \gamma_{ijk}u^i v^j w^k,$$

$$\text{Third term} : \underbrace{\gamma_{ijk}}_{=\gamma_{kij}} u^k v^i w^j = \gamma_{kij}u^k v^i w^j = \underbrace{\gamma_{kij}u^k v^i w^j}_{k\to i, i\to j, j\to k} = \gamma_{ijk}u^i v^j w^k,$$

$$\text{Forth term} : \underbrace{-\gamma_{ijk}}_{+\gamma_{jik}} u^j v^i w^k = \gamma_{jik}u^j v^i w^k = \underbrace{\gamma_{jik}u^j v^i w^k}_{j\to i, i\to j} = \gamma_{ijk}u^i v^j w^k,$$

$$\text{Fifth term} : \underbrace{-\gamma_{ijk}}_{+\gamma_{kji}} u^k v^j w^i = \gamma_{kji}u^k v^j w^i = \underbrace{\gamma_{kji}u^k v^j w^i}_{k\to i, i\to k} = \gamma_{ijk}u^i v^j w^k,$$

$$\text{Sixth term} : \underbrace{-\gamma_{ijk}}_{+\gamma_{ikj}} u^i v^k w^j = \gamma_{ikj}u^i v^k w^j = \underbrace{\gamma_{ikj}u^i v^k w^j}_{k\to j, j\to k} = \gamma_{ijk}u^i v^j w^k.$$

We add the summands , $= 6\,\gamma_{[ijk]}u^i v^j w^k$, and get for the complete expression:

$$\tilde{\gamma}\left(\vec{u}, \vec{v}, \vec{w}\right) = \frac{1}{3!} \cdot 6\,\gamma_{[ijk]}u^i v^j w^k = \gamma_{[ijk]}u^i v^j w^k.$$

If we substitute the Levi-Civita tensor ϵ_{ijk} for $\gamma_{[ijk]}$, which we are allowed to do because of its complete antisymmetry, among other permutation properties, then the expression gives us the determinant of the three vectors of an $n = 3$ vector space:

$$\epsilon_{ijk}u^i v^j w^k = \left|\vec{u}\ \ \vec{v}\ \ \vec{w}\right| = \begin{vmatrix} u^1 & v^1 & w^1 \\ u^2 & v^2 & w^2 \\ u^3 & v^3 & w^3 \end{vmatrix}.$$

8.3   Integration revisited

What is the area A enclosed by two vectors $\vec{u}$ and $\vec{v}$ in a plane? From vector algebra we know the answer that the determinant of the vector matrix is to be determined. Ignoring the sign, we get:

$$A = \begin{vmatrix} u^1 & v^1 \\ u^2 & v^2 \end{vmatrix}.$$

Now let us reduce the vectors to infinitesimal size $\vec{u} \to d\vec{u}$ and $\vec{v} \to d\vec{v}$. And the enclosed area dA now results accordingly in:

$$dA = \begin{vmatrix} du^1 & dv^1 \\ du^2 & dv^2 \end{vmatrix} = du^1 dv^2 - dv^1 du^2.$$

In this result we recognise the definition of the wedge product equation (8.5a):

$$d\tilde{x}^i \wedge d\tilde{x}^j = d\tilde{x}^i \otimes d\tilde{x}^j - d\tilde{x}^j \otimes d\tilde{x}^i = d\tilde{x}^i d\tilde{x}^j - d\tilde{x}^j d\tilde{x}^i.$$

Therefore, the statement that *the integration over an n-dimensional region of a manifold is correct only by use of an n-form* can be well understood. Compare this with the discussion in Section 8.11.

Box 8.4
Jacobian revisited

We consider the 2-dimensional case of an infinitesimal surface dA and a coordinate transformation from $\left\{ x^{i'} \right\}$ to $\left\{ x^i \right\}$. The x^i-coordinates are related to the $x^{i'}$-coordinates by the transformation

$$x^1 = u(x^{1'}, x^{2'}),$$
$$x^2 = v(x^{1'}, x^{2'}).$$

And by the chain rule we get

$$dx^1 = \frac{\partial u}{\partial x^{1'}} dx^{1'} + \frac{\partial u}{\partial x^{2'}} dx^{2'},$$
$$dx^2 = \frac{\partial v}{\partial x^{1'}} dx^{1'} + \frac{\partial v}{\partial x^{2'}} dx^{2'}.$$

And for dA we obtain

$$dA = dx^1 \wedge dx^2$$

$$= \left(\frac{\partial u}{\partial x^{1'}} dx^{1'} + \frac{\partial u}{\partial x^{2'}} dx^{2'} \right) \wedge \left(\frac{\partial v}{\partial x^{1'}} dx^{1'} + \frac{\partial v}{\partial x^{2'}} dx^{2'} \right)$$

$$= \left(\frac{\partial u}{\partial x^{1'}} \frac{\partial v}{\partial x^{2'}} - \frac{\partial u}{\partial x^{2'}} \frac{\partial v}{\partial x^{1'}} \right) dx^{1'} \wedge dx^{2'}$$

$$= \left(\frac{\partial x^1}{\partial x^{1'}} \frac{\partial x^2}{\partial x^{2'}} - \frac{\partial x^1}{\partial x^{2'}} \frac{\partial x^2}{\partial x^{1'}} \right) dx^{1'} \wedge dx^{2'}$$

$$= \det \begin{pmatrix} \frac{\partial x^1}{\partial x^{1'}} & \frac{\partial x^1}{\partial x^{2'}} \\ \frac{\partial x^2}{\partial x^{1'}} & \frac{\partial x^2}{\partial x^{2'}} \end{pmatrix} dx^{1'} \wedge dx^{2'} \, ,$$

with $J = \det \begin{pmatrix} \frac{\partial x^1}{\partial x^{1'}} & \frac{\partial x^1}{\partial x^{2'}} \\ \frac{\partial x^2}{\partial x^{1'}} & \frac{\partial x^2}{\partial x^{2'}} \end{pmatrix}$ as the Jacobian of the transformation that keeps the area element dA invariant in both coordinate systems:

$$dx^1 dx^2 = J dx^{1'} dx^{2'} \, .$$

We want to briefly and compactly show the way to the integral of a form. An n-form

$$d\tilde{x}^1 \wedge \cdots \wedge d\tilde{x}^n \tag{8.10}$$

of an n-dimensional vector space is an n-dimensional parallelepiped. It has as components those of the Levi-Civita symbol. Box 8.5 shows this for a 3-form.

$$d\tilde{x}^1 \wedge \cdots \wedge d\tilde{x}^n = \frac{1}{n!} \epsilon_{i_1 \dots i_n} d\tilde{x}^{i_1} \wedge \cdots \wedge d\tilde{x}^{i_n}. \tag{8.11}$$

The Levi-Civita symbol is a tensor density, Section 4.6. And hence an n-form is not a tensor but a tensor density.

Box 8.5

The components of $d\tilde{x}^1 \wedge d\tilde{x}^2 \wedge d\tilde{x}^3$ on a three dimensional manifold are determined by processing the basis vectors $\{\vec{e}_i, \vec{e}_j, \vec{e}_k\}$.

$$
\begin{aligned}
d\tilde{x}^1 &\wedge d\tilde{x}^2 \wedge d\tilde{x}^3 \, (\vec{e}_i, \vec{e}_j, \vec{e}_k) \\
&= d\tilde{x}^1(\vec{e}_i)d\tilde{x}^2(\vec{e}_j)d\tilde{x}^3(\vec{e}_k) \\
&+ d\tilde{x}^2(\vec{e}_i)d\tilde{x}^3(\vec{e}_j)d\tilde{x}^1(\vec{e}_k) \\
&+ d\tilde{x}^3(\vec{e}_i)d\tilde{x}^1(\vec{e}_j)d\tilde{x}^2(\vec{e}_k) \\
&- d\tilde{x}^2(\vec{e}_i)d\tilde{x}^1(\vec{e}_j)d\tilde{x}^3(\vec{e}_k) \\
&- d\tilde{x}^3(\vec{e}_i)d\tilde{x}^2(\vec{e}_j)d\tilde{x}^1(\vec{e}_k) \\
&- d\tilde{x}^1(\vec{e}_i)d\tilde{x}^3(\vec{e}_j)d\tilde{x}^2(\vec{e}_k) \\
&= \delta_i^1\delta_j^2\delta_k^3 + \delta_i^2\delta_j^3\delta_k^1 + \delta_i^3\delta_j^1\delta_k^2 - \delta_i^2\delta_j^1\delta_k^3 - \delta_i^3\delta_j^2\delta_k^1 - \delta_i^1\delta_j^3\delta_k^2 \\
&\equiv \epsilon_{ijk}.
\end{aligned}
$$

For definition of Levi-Civita symbol ϵ_{ijk} see equation (4.37).

To study the transformation behaviour of the components of an n-form, we need to consider the transformation behaviour of the Levi-Civita symbol $\epsilon_{i_1 \dots i_n}$. We have already done this in Section 4.6, equation (4.40). Together with equation (4.12) we get the relation

$$
\sqrt{|\det g'|}\, \epsilon'_{i_1 \dots i_n} = \sqrt{|\det g|}\, \epsilon_{i_1 \dots i_n}. \tag{8.12}
$$

We define a new object for the components of an n-form respecting the choice of coordinates by the metric determinant

$$
\omega_{i_1 \dots i_n} = \sqrt{|\det g|}\, \epsilon_{i_1 \dots i_n}, \tag{8.13}
$$

and obtain for $\omega_{i_1 \dots i_n}$ the transformation

$$
\begin{aligned}
\omega'_{j_1 \dots j_n} &= \omega_{i_1 \dots i_n} F_{j_1}^{i_1} \cdots F_{j_n}^{i_n} = \sqrt{|\det g|}\, \epsilon_{i_1 \dots i_n} F_{j_1}^{i_1} \cdots F_{j_n}^{i_n} \\
&= \sqrt{\frac{|\det g'|}{(\det F)^2}}\, (\det F)\, \epsilon_{j_1 \dots j_n} = \operatorname{sign}(\det F)\sqrt{|\det g'|}\, \epsilon_{j_1 \dots j_n}.
\end{aligned}
$$

$\omega'_{j_1 \dots j_n}$ depends on the handedness (see Section 8.4) of the coordinate transformation F. We choose a positive handedness and receive:

$$
\omega'_{j_1 \dots j_n} = \sqrt{|\det g'|}\, \epsilon_{j_1 \dots j_n}, \text{ or } \omega_{i_1 \dots i_n} = \sqrt{|\det g|}\, \epsilon_{i_1 \dots i_n}. \tag{8.14}
$$

We complete the n-form $\tilde{\omega}$ from equation (8.11) with the scale factor $\sqrt{|\det g|}$ of the chosen coordinates and obtain the following expression

$$\tilde{\omega} = \sqrt{|\det g|}\, d\tilde{x}^1 \wedge \cdots \wedge d\tilde{x}^n \tag{8.15a}$$

$$= \frac{1}{n!} \sqrt{|\det g|}\, \epsilon_{i_1 \ldots i_n}\, d\tilde{x}^{i_1} \wedge \cdots \wedge d\tilde{x}^{i_n}$$

$$= \frac{1}{n!}\, \omega_{i_1 \ldots i_n}\, d\tilde{x}^{i_1} \wedge \cdots \wedge d\tilde{x}^{i_n}. \tag{8.15b}$$

$\tilde{\omega}$ is also still referred to as *volume form.*

Summing the volume n-form $\tilde{\omega}$ over a uniquely oriented region R of an n-dimensional manifold M, and letting the size of each volume cell approach zero, the result can be represented as an integral:

$$\sum_R \tilde{\omega} \longrightarrow \int_R \tilde{\omega} = \int_R \sqrt{|\det g|}\, dx^1 \ldots dx^n. \tag{8.16}$$

The middle expression of equation (8.16) is a new writing for an integral on a manifold and the expression to the right of it is the ordinary integral from calculus. We can still " weighted " the volume form with a smooth, scalar function $f\left(x^1,, \ldots, x^n\right)$ such that

$$\tilde{\omega} = f \sqrt{|\det g|}\, d\tilde{x}^1 \wedge \cdots \wedge d\tilde{x}^n$$

and thus obtain an integration of this function f over the selected Region R:

$$\int_R \tilde{\omega} = \int_R f\left(x^1,, \ldots, x^n\right) \sqrt{|\det g|}\, dx^1 \ldots dx^n. \tag{8.17}$$

8.4   Oriented vector space

In Section 8.3 we introduced the volume form $\tilde{\omega}$ of an n-dimensional manifold M whose coordinates are $\left\{x^1, \ldots x^n\right\}$. It is an n-form, a one-dimensional dual vector space, equation (8.3). We choose a basis $\left\{\partial_{x^1}, \ldots, \partial_{x^1}\right\} \equiv \left\{\vec{e}_1, \ldots, \vec{e}_n\right\}$ at a point $\mathcal{P}$ of the manifold. Since the basis is linearly independent, $\tilde{\omega}\left(\vec{e}_{i_1}, \ldots, \vec{e}_{i_n}\right)$ is a non-zero number. This is therefore either positive or negative. Hence $\tilde{\omega}$ separates the basis of a manifold into two *classes*. A positive number is called *right-handed*, a negative number is called *left-handed*.

A manifold is called *orientable*, if one can define a basis $\tilde{\omega}\left(\vec{e}_{i_1}(\mathcal{P}), \ldots, \vec{e}_{i_n}(\mathcal{P})\right)$, which leads to a consistent *handedness* for all points $\mathcal{P}$ of the manifold.

Example 8.3

For a $n = 3$ vector space we want to determine the different Cartesian bases that lead to right-handedness and left-handedness, respectively. We label the coordinate axes $\{x, y, z\}$ with the numbers $\{1, 2, 3\}$. There are $3! = 6$ different bases. The first line shows an even permutation of the initial number series $\{1, 2, 3\}$, the second line an odd permutation of it:

$$\{\vec{e}_1, \vec{e}_2, \vec{e}_3\}, \ \{\vec{e}_2, \vec{e}_3, \vec{e}_1\}, \ \{\vec{e}_3, \vec{e}_1, \vec{e}_2\},$$
$$\text{and}$$
$$\{\vec{e}_2, \vec{e}_1, \vec{e}_3\}, \ \{\vec{e}_3, \vec{e}_2, \vec{e}_1\}, \ \{\vec{e}_1, \vec{e}_3, \vec{e}_2\}.$$

We now let the volume form $\tilde{\omega} = \frac{1}{3!}\sqrt{|\det g|}\, \epsilon_{i_1 \, \epsilon_{i_2} \, i_3}\, d\tilde{x}^{i_1} \wedge d\tilde{x}^{i_2} \wedge d\tilde{x}^{i_3}$ with $\det g = 1$ process the individual basis sets. To shorten the calculations we use the result of Box 8.3:

$$\tilde{\omega}\left(\vec{e}_1, \vec{e}_2, \vec{e}_3\right) = \begin{vmatrix} 1 & 0 & 0 \\ 0 & 1 & 0 \\ 0 & 0 & 1 \end{vmatrix} = 1, \quad \tilde{\omega}\left(\vec{e}_2, \vec{e}_3, \vec{e}_1\right) = \begin{vmatrix} 0 & 0 & 1 \\ 1 & 0 & 0 \\ 0 & 1 & 0 \end{vmatrix} = 1,$$

$$\tilde{\omega}\left(\vec{e}_3, \vec{e}_1, \vec{e}_2\right) = \begin{vmatrix} 0 & 1 & 0 \\ 0 & 0 & 1 \\ 1 & 0 & 0 \end{vmatrix} = 1, \quad \tilde{\omega}\left(\vec{e}_2, \vec{e}_1, \vec{e}_3\right) = \begin{vmatrix} 0 & 1 & 0 \\ 1 & 0 & 0 \\ 0 & 0 & 1 \end{vmatrix} = -1,$$

$$\tilde{\omega}\left(\vec{e}_3, \vec{e}_2, \vec{e}_1\right) = \begin{vmatrix} 0 & 0 & 1 \\ 0 & 1 & 0 \\ 1 & 0 & 0 \end{vmatrix} = -1, \quad \tilde{\omega}\left(\vec{e}_1, \vec{e}_3, \vec{e}_2\right) = \begin{vmatrix} 1 & 0 & 0 \\ 0 & 0 & 1 \\ 0 & 1 & 0 \end{vmatrix} = -1.$$

$\blacksquare$

8.5 p-vectors

The Grassmann algebra we set out in Section 8.2 for completely antisymmetric $(0, p)$-tensors, the p-forms, can be extended in the same way for $(p, 0)$ tensors.

A completely antisymmetric $(p,0)$-tensor is called a p-vector. The corresponding vector space is denoted by $\wedge^p \mathcal{V}$. The number of free components of a p-vector in an n-dimensional vector space is also $C_p^n = \frac{n!}{p!(n-p)!} = \binom{n}{p}$, equation (8.3), because of the algebra identical to a p-form.

A 1-vector is the well-known vector of a vector space $\wedge^1 \mathcal{V} \equiv \mathcal{V}$. We conform the notation of the basis vectors of the p-vectors to that of the p-forms, as directional differentials

$$\vec{e}_i \equiv d\vec{x}_i.$$

A 1-vector is thus notated as $\vec{v} = v^i d\vec{x}_i$, and for $n = 3 : \vec{v} = v^1 d\vec{x}_1 + v^2 d\vec{x}_2 + v^3 d\vec{x}_3$. As a straightforward example, we want to form on an $n = 3$ vector space a 2-vector from two 1-vectors. Let $\vec{u} = u^i d\vec{x}_i = u^1 d\vec{x}_1 + u^2 d\vec{x}_2 + u^3 d\vec{x}_3$ and $\vec{v} = v^i d\vec{x}_i = v^1 d\vec{x}_1 + v^2 d\vec{x}_2 + v^3 d\vec{x}_3$, then

$$
\begin{aligned}
\vec{u} \wedge \vec{v} &= \left(u^i d\vec{x}_i\right) \wedge \left(v^j d\vec{x}_j\right) = u^i v^j d\vec{x}_i \wedge d\vec{x}_j \\
&= u^1 v^2 d\vec{x}_1 \wedge d\vec{x}_2 + u^1 v^3 d\vec{x}_1 \wedge d\vec{x}_3 + u^2 v^3 d\vec{x}_2 \wedge d\vec{x}_3 + \\
&\quad + u^2 v^1 d\vec{x}_2 \wedge d\vec{x}_1 + u^3 v^1 d\vec{x}_3 \wedge d\vec{x}_1 + u^3 v^2 d\vec{x}_3 \wedge d\vec{x}_2 \\
&= \left(u^2 v^3 - u^3 v^2\right) d\vec{x}_2 \wedge d\vec{x}_3 + \\
&\quad + \left(u^3 v^1 - u^1 v^3\right) d\vec{x}_3 \wedge d\vec{x}_1 + \\
&\quad + \left(u^1 v^2 - u^2 v^1\right) d\vec{x}_1 \wedge d\vec{x}_2 .
\end{aligned}
\tag{8.18}
$$

The order of the summation terms was chosen intentionally so that the expression is consistent with the comments on Examples 8.4 and 8.5 . Note the cyclic order of the indices.

Example 8.4

We know that the vector $\vec{u} \times \vec{v}$ as the result of a cross product of two vectors $\vec{u}$ and $\vec{v}$ has different properties than the starting vectors $\vec{u}$ and $\vec{v}$. $\vec{u} \times \vec{v}$ has the dimension of an area, unlike the vectors $\vec{u}$ and $\vec{v}$, whose dimension is a length. $\vec{u} \times \vec{v}$ also has other symmetry properties than $\vec{u}$ and $\vec{v}$: $\vec{u}$ and $\vec{v}$ change the sign in case of spatial reflection, $\vec{u} \times \vec{v}$ does not (axial vector). Based on these properties of the cross product $\vec{u} \times \vec{v}$, it is therefore not unexpected that its components $\left(u^2 v^3 - u^3 v^2, u^3 v^1 - u^1 v^3, u^1 v^2 - u^2 v^1\right)$ coincide with the wedge product $\vec{u} \wedge \vec{v}$, equation (8.18). Compare supplementary Example 8.5.

■

8.6 Dual spaces of p-forms and p-vectors

As we have learned in the previous sections, the vector spaces of p-forms and p-vectors at any point $\mathcal{P}$ in an n-dimensional manifold M each have a dimension of C_p^n, (8.3). Thus, at each point $\mathcal{P}$ of an n-dimensional manifold M there are *four* vector spaces: the vector spaces of p-forms $\wedge^p \mathcal{V}^\star$, $(n-p)$-forms $\wedge^{n-p}\mathcal{V}^\star$, p-vectors $\wedge^p \mathcal{V}$ and $(n-p)$-vectors $\wedge^{n-p}\mathcal{V}$. All these four vector spaces have the **same dimension**, namely $C_p^n = C_{n-p}^n$, Figure 8.1.

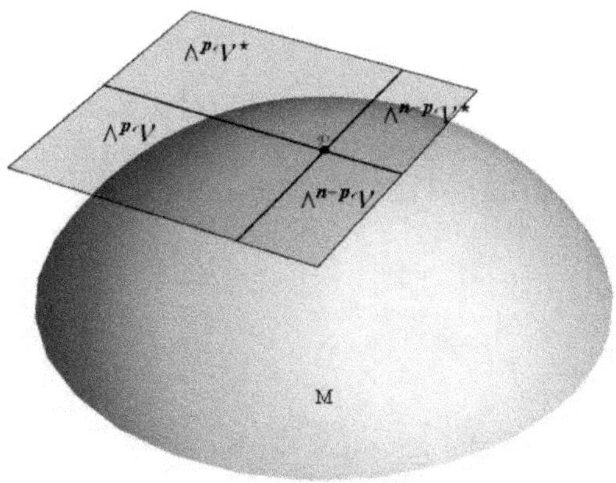

Figure 8.1: The four vector spaces $\wedge^p \mathcal{V}^\star$, $\wedge^{n-p}\mathcal{V}^\star$, $\wedge^p \mathcal{V}$ and $\wedge^{n-p}\mathcal{V}$ at point $\mathcal{P}$ on a manifold M.

Thus a $1-1$ mapping exists between these vector spaces. p-forms can be transformed into $(n-p)$-vectors and vice versa via *dual mapping*, Section 8.8. p-forms can be transformed into p-vectors and vice versa using the *metric*, which we already know from $(0,p)$-tensors and $(p,0)$-tensors, Section 4.11. The *Hodge dual mapping* of a p-form to an $(n-p)$-form, or p-vector to an $(n-p)$-vector is done by the so-called Hodge operator $*$ or star-operator, Section 8.7. In equation (8.19), these mentioned mappings are symbolically compiled.

$$\wedge^p \mathcal{V}^\star \xleftrightarrow{\tilde{\omega}} \wedge^{n-p} \mathcal{V},$$
$$\wedge^{n-p} \mathcal{V}^\star \xleftrightarrow{\tilde{\omega}} \wedge^p \mathcal{V},$$
$$\wedge^p \mathcal{V}^\star \xleftrightarrow{g,\,g} \wedge^p \mathcal{V}, \qquad (8.19)$$
$$\wedge^p \mathcal{V}^\star \xleftrightarrow{*} \wedge^{n-p} \mathcal{V}^\star,$$
$$\wedge^p \mathcal{V} \xleftrightarrow{*} \wedge^{n-p} \mathcal{V}.$$

8.7 Hodge dual operator $*$

In this section we use an orthonormal basis of Euclidean ($n = 3$) or Minkowski ($n = 4$) space , Section 4.4, with defined orientation, Section 8.4. Thereby we use the notation of the basis vectors as before, i.e.:

$$d\hat{\tilde{x}}^i \equiv d\tilde{x}^i, \quad d\hat{\vec{x}}_i \equiv d\vec{x}_i.$$

Because of the equality of the dimensions of the p-form vector space $\wedge^p \mathcal{V}^\star$ with the $(n - p)$-form vector space $\wedge^{n-p} \mathcal{V}^\star$, $C_p^n = C_{n-p}^n$, you can assign an $(n - p)$-form to each p-form $\tilde{p}$, which is marked $*\tilde{p}$. This assignment rule is called *Hodge dual* and is marked with a star $*$:

$$* : \wedge^p \mathcal{V}^\star \longrightarrow \wedge^{n-p} \mathcal{V}^\star. \qquad (8.20)$$

And the same is true for p-vectors:

$$* : \wedge^p \mathcal{V} \longrightarrow \wedge^{n-p} \mathcal{V}. \qquad (8.21)$$

The Hodge operator $*$ acts only on forms or vectors but not on scalars:

$$* \left(f \, d\tilde{x}^i \right) = f * d\tilde{x}^i,$$
$$* \left(f \, d\vec{x}_i \right) = f * d\vec{x}_i,$$
$$*\tilde{p} = * \left(p_i \, d\tilde{x}^i \right) = p_i * d\tilde{x}^i, \qquad (8.22)$$
$$*\vec{v} = * \left(v^i \, d\vec{x}_i \right) = v^i * d\vec{x}_i.$$

For a Euclidean dual vector space $\mathcal{V}^\star$ with its exterior algebra $\wedge^p \mathcal{V}^\star$, the following holds: Let $d\tilde{x}^I$ be the basis of a p-form, (8.7). The Hodge dual basis $\left(*d\tilde{x}^I\right)$ is the *complement* of the n-form $\tilde{\omega}$ (equation (8.15a) with $|\det g| = 1$):

$$*d\tilde{x}^I = *\left(d\tilde{x}^{i_1} \wedge \ldots \wedge d\tilde{x}^{i_p}\right) = \pm \left\{d\tilde{x}^{j_1} \wedge \ldots \wedge d\tilde{x}^{j_{n-p}}\right\}.$$

If $\{i_1, \ldots i_p, j_1 \ldots j_{n-p}\}$ is an even permutation of $\{1, \ldots n\}$, then the '+' sign applies, and the '-' sign applies for an odd permutation.

For example: In a 3-dimensional Euclidean vector space the Hodge operator $*$ assigns every 2-form to 1-form and vice versa, every 3-form to a 0-form and vice versa. Note: A cyclic permutation leaves an even permutation even.

$$*d\tilde{x}^1 = d\tilde{x}^2 \wedge d\tilde{x}^3,$$
$$\text{and cyclically}: *d\tilde{x}^2 = d\tilde{x}^3 \wedge d\tilde{x}^1, \quad *d\tilde{x}^3 = d\tilde{x}^1 \wedge d\tilde{x}^2,$$
$$*\left(d\tilde{x}^2 \wedge d\tilde{x}^3\right) = d\tilde{x}^1,$$
$$\text{and cyclically}: *\left(d\tilde{x}^3 \wedge d\tilde{x}^1\right) = d\tilde{x}^2, \tag{8.23}$$
$$*\left(d\tilde{x}^1 \wedge d\tilde{x}^2\right) = d\tilde{x}^3,$$
$$*\left(d\tilde{x}^1 \wedge d\tilde{x}^2 \wedge d\tilde{x}^3\right) = 1, \; 0 - \text{form}.$$

Example 8.5 continuation of Example 8.4

The exterior product of two vectors $\vec{u}$ and $\vec{v}$, see equation (8.18), is

$$\vec{u} \wedge \vec{v} = \left(u^2 v^3 - u^3 v^2\right) d\vec{x}_2 \wedge d\vec{x}_3 +$$
$$+ \left(u^3 v^1 - u^1 v^3\right) d\vec{x}_3 \wedge d\vec{x}_1 +$$
$$+ \left(u^1 v^2 - u^2 v^1\right) d\vec{x}_1 \wedge d\vec{x}_2.$$

The Hodge dual of $\vec{u} \wedge \vec{v} \in \wedge^2 \mathcal{V}$ is

$$*\left(\vec{u} \wedge \vec{v}\right) = \left(u^2 v^3 - u^3 v^2\right) * \left(d\vec{x}_2 \wedge d\vec{x}_3\right) +$$
$$+ \left(u^3 v^1 - u^1 v^3\right) * \left(d\vec{x}_3 \wedge d\vec{x}_1\right) +$$
$$+ \left(u^1 v^2 - u^2 v^1\right) * \left(d\vec{x}_1 \wedge d\vec{x}_2\right)$$
$$= \left(u^2 v^3 - u^3 v^2\right) d\vec{x}_1 + \left(u^3 v^1 - u^1 v^3\right) d\vec{x}_2 + \left(u^1 v^2 - u^2 v^1\right) d\vec{x}_3.$$

The result is the cross product of two vectors,

$$*\left(\vec{u} \wedge \vec{v}\right) = \vec{u} \times \vec{v}. \quad \blacksquare$$

For two p-forms $\tilde{p}, \tilde{q} \in \wedge^p \mathcal{V}^\star$ then holds

$$\tilde{p} \wedge (*\tilde{q}) = \boldsymbol{g}\,(\tilde{p}, \tilde{q})\,\tilde{\omega}, \tag{8.24}$$

with $\boldsymbol{g}$ as the dual metric. Proof in Box 8.6.

Box 8.6

$$\tilde{p} = p_I d\tilde{x}^I \quad \text{and} \quad \tilde{q} = q_J d\tilde{x}^J,$$
$$\tilde{p} \wedge *\tilde{q} = p_I d\tilde{x}^I \wedge *q_J d\tilde{x}^J = p_I q_J d\tilde{x}^I \wedge *d\tilde{x}^J.$$

The wedge product $d\tilde{x}^I \wedge *d\tilde{x}^J$ disappears unless I and J are equal. Therefore applies $d\tilde{x}^I \wedge *d\tilde{x}^I = \tilde{\omega}$, $p_I q_I = \boldsymbol{g}\,(\tilde{p}, \tilde{q})$. And thus

$$\tilde{p} \wedge (*\tilde{q}) = \boldsymbol{g}\,(\tilde{p}, \tilde{q})\,\tilde{\omega}.$$

Equation (8.24) also applies to metrics with a signature $s \neq 0$.

As we saw in Section 4.4, the canonical metric tensor contains only ± 1. The adjusted basis is orthonormal. If $d\tilde{x}^i$ is the orthonormal basis of the 1-form vector space $\wedge^1 \mathcal{V}^\star$ and $\boldsymbol{g}$ the dual metric tensor, then

$$\boldsymbol{g}\left(d\tilde{x}^i, d\tilde{x}^j\right) = g^{ij} = \pm \delta^{ij}. \tag{8.25}$$

Therefore, for an n-form $\tilde{\omega}$ of an n-form vector space $\wedge^n \mathcal{V}^\star$, ($|\det g| = 1$, signature s) applies:

$$\boldsymbol{g}\,(\tilde{\omega}, \tilde{\omega}) = \boldsymbol{g}\left(d\tilde{x}^1, d\tilde{x}^1\right) \cdots \boldsymbol{g}\left(d\tilde{x}^n, d\tilde{x}^n\right) = (-1)^s. \tag{8.26}$$

Remark 8.1

The facts mentioned above and the considerations in the following sections apply equally to a p-vector space $\wedge^p \mathcal{V}$. Please note that the metric tensor $\mathbf{g}(d\vec{x}_i, d\vec{x}_j) = g_{ij}$ must then be used.

$\blacksquare$

Because of the signature s we have to consider the following when forming $*\tilde{\omega}$ and $*1$, using equations (8.24) and (8.26):

$$\tilde{\omega} \wedge (*\tilde{\omega}) = \boldsymbol{g}(\tilde{\omega}, \tilde{\omega})\,\tilde{\omega} = (-1)^s\,\tilde{\omega},$$
$$*\tilde{\omega} = (-1)^s.$$
$$*1 = 1 \wedge *1 = \boldsymbol{g}(1,1)\,\tilde{\omega} = \tilde{\omega}.$$
$$*(*1) = *\tilde{\omega} = (-1)^s. \tag{8.27}$$

We can express the *inner product* $\langle \tilde{p}, \tilde{q} \rangle$ between two forms $\tilde{p}, \tilde{q} \in \wedge^p V^\star$ by applying equation (8.24):

$$*\left(\tilde{p} \wedge (*\tilde{q})\right) = *\boldsymbol{g}(\tilde{p}, \tilde{q})\,\tilde{\omega} = \boldsymbol{g}(\tilde{p}, \tilde{q})\,*\tilde{\omega} = \boldsymbol{g}(\tilde{p}, \tilde{q})\,(-1)^s,$$
$$\Longrightarrow \tag{8.28}$$
$$\langle \tilde{p}, \tilde{q} \rangle = \tilde{p} \cdot \tilde{q} = \boldsymbol{g}(\tilde{p}, \tilde{q}) = (-1)^s * \left(\tilde{p} \wedge (*\tilde{q})\right).$$

This definition is valid for any dimension and signature.

Hodge operations $*$ often results in an expression $**$. Which term do we need to assign to $**$? Starting from an n-dimensional vector space we consider a p-form $\in \wedge^p V^\star$ with the orthonormal basis $d\tilde{x}^I$. Using equation (8.24):

$$d\tilde{x}^I \wedge \left(*d\tilde{x}^I\right) = \boldsymbol{g}\left(d\tilde{x}^I, d\tilde{x}^I\right)\tilde{\omega}$$
$$= \boldsymbol{g}\left(d\tilde{x}^I, d\tilde{x}^I\right) d\tilde{x}^I \wedge d\tilde{x}^J, \tag{8.29}$$

with $d\tilde{x}^J$ as the complementary basis set to $d\tilde{x}^I$. Therefore

$$*d\tilde{x}^I = \boldsymbol{g}\left(d\tilde{x}^I, d\tilde{x}^I\right) d\tilde{x}^J. \tag{8.30}$$

For the product $d\tilde{x}^I \wedge d\tilde{x}^J$ we get the following expression after shifting the individual basis elements through the product:

$$d\tilde{x}^I \wedge d\tilde{x}^J = (-1)^{p(n-p)} d\tilde{x}^J \wedge d\tilde{x}^I. \tag{8.31}$$

And similar to (8.29) with the result of (8.30) we get

$$d\tilde{x}^J \wedge \left(*d\tilde{x}^J\right) = \boldsymbol{g}\left(d\tilde{x}^J, d\tilde{x}^J\right)\tilde{\omega}$$
$$= \boldsymbol{g}\left(d\tilde{x}^J, d\tilde{x}^J\right) d\tilde{x}^I \wedge d\tilde{x}^J$$
$$= \boldsymbol{g}\left(d\tilde{x}^J, d\tilde{x}^J\right)(-1)^{p(n-p)} d\tilde{x}^J \wedge d\tilde{x}^I, \tag{8.32}$$
$$\Longrightarrow$$
$$*d\tilde{x}^J = \boldsymbol{g}\left(d\tilde{x}^J, d\tilde{x}^J\right)(-1)^{p(n-p)} d\tilde{x}^I.$$

And now we combine the results and get the following general expression:

$$
\begin{aligned}
* * d\tilde{x}^I &= * \left(\boldsymbol{g} \left(d\tilde{x}^I, d\tilde{x}^I \right) d\tilde{x}^J \right) = \boldsymbol{g} \left(d\tilde{x}^I, d\tilde{x}^I \right) * d\tilde{x}^J \\
&= \boldsymbol{g} \left(d\tilde{x}^I, d\tilde{x}^I \right) \left(\boldsymbol{g} \left(d\tilde{x}^J, d\tilde{x}^J \right) (-1)^{p(n-p)} d\tilde{x}^I \right) \\
&= (-1)^{p(n-p)} \boldsymbol{g} \left(d\tilde{x}^I, d\tilde{x}^I \right) \boldsymbol{g} \left(d\tilde{x}^J, d\tilde{x}^J \right) d\tilde{x}^I \\
&= (-1)^{p(n-p)} \boldsymbol{g} \left(\tilde{\omega}, \tilde{\omega} \right) d\tilde{x}^I \\
&= (-1)^{p(n-p)+s} d\tilde{x}^I .
\end{aligned}
\tag{8.33}
$$

For a p-form $\tilde{p} = p_I d\tilde{x}^I$:

$$
* * \tilde{p} = (-1)^{p(n-p)+s} \, \tilde{p}.
\tag{8.34}
$$

Box 8.7 Hodge operator $*$ in Minkowski space

We choose the representation for p-vectors$\in \mathcal{V}$ rather than p-forms. With the Minkowski metric η holds:

$$
\begin{aligned}
\eta \left(d\vec{x}_0, d\vec{x}_0 \right) &= \eta \left(dt, dt \right) = -1, \\
\eta \left(d\vec{x}_1, d\vec{x}_1 \right) &= \eta \left(d\vec{x}_2, d\vec{x}_2 \right) = \eta \left(d\vec{x}_3, d\vec{x}_3 \right) = 1.
\end{aligned}
$$

According to $C_p^n = \binom{n}{p}$, (8.3): four basis vectors $d\vec{x}_i \in \wedge^1 \mathcal{V}$, six basis vectors $d\vec{x}_i \wedge d\vec{x}_j \in \wedge^2 \mathcal{V}$ and four basis vectors $d\vec{x}_i \wedge d\vec{x}_j \wedge d\vec{x}_k \in \wedge^3 \mathcal{V}$.

We want to calculate exemplarily the Hodge dual of the 1-vectors, in order to be able to determine afterwards the scalar product of two 4-event vectors. Respecting Remark 8.1 and using (8.24) for p-vectors:

$$d\vec{x}_0 \wedge (*d\vec{x}_0) = \eta\,(d\vec{x}_0, d\vec{x}_0)\,\vec{\omega} = -d\vec{x}_0 \wedge d\vec{x}_1 \wedge d\vec{x}_2 \wedge d\vec{x}_3,$$
$$*d\vec{x}_0 = -d\vec{x}_1 \wedge d\vec{x}_2 \wedge d\vec{x}_3.$$
$$d\vec{x}_1 \wedge (*d\vec{x}_1) = \eta\,(d\vec{x}_1, d\vec{x}_1)\,\vec{\omega} = d\vec{x}_0 \wedge d\vec{x}_1 \wedge d\vec{x}_2 \wedge d\vec{x}_3,$$
$$*d\vec{x}_1 = -d\vec{x}_0 \wedge d\vec{x}_2 \wedge d\vec{x}_3.$$
$$d\vec{x}_2 \wedge (*d\vec{x}_2) = \eta\,(d\vec{x}_2, d\vec{x}_2)\,\vec{\omega} = d\vec{x}_0 \wedge d\vec{x}_1 \wedge d\vec{x}_2 \wedge d\vec{x}_3,$$
$$*d\vec{x}_2 = d\vec{x}_0 \wedge d\vec{x}_1 \wedge d\vec{x}_3.$$
$$d\vec{x}_3 \wedge (*d\vec{x}_3) = \eta\,(d\vec{x}_3, d\vec{x}_3)\,\vec{\omega} = d\vec{x}_0 \wedge d\vec{x}_1 \wedge d\vec{x}_2 \wedge d\vec{x}_3,$$
$$*d\vec{x}_3 = -d\vec{x}_0 \wedge d\vec{x}_1 \wedge d\vec{x}_2.$$

Box 8.8

Determination of the scalar product of two 4-event vectors in Minkowski space with equation (8.28)

The two 4-event be:

$$\vec{u} = u^0 d\vec{x}_0 + u^1 d\vec{x}_1 + u^2 d\vec{x}_2 + u^3 d\vec{x}_3,$$
$$\vec{v} = v^0 d\vec{x}_0 + v^1 d\vec{x}_1 + v^2 d\vec{x}_2 + v^3 d\vec{x}_3.$$

Determination of $*\vec{v}$ with the results of Box 8.7.

$$*\vec{v} = v^0 * d\vec{x}_0 + v^1 * d\vec{x}_1 + v^{2}{*}d\vec{x}_2 + v^3 * d\vec{x}_3$$
$$= -v^0 d\vec{x}_1 \wedge d\vec{x}_2 \wedge d\vec{x}_3 - v^1 d\vec{x}_0 \wedge d\vec{x}_2 \wedge d\vec{x}_3 +$$
$$+ v^2 d\vec{x}_0 \wedge d\vec{x}_1 \wedge d\vec{x}_3 - v^3 d\vec{x}_0 \wedge d\vec{x}_1 \wedge d\vec{x}_2.$$

And considering $d\vec{x}_i \wedge d\vec{x}_i = 0$:

$$\vec{u} \wedge (*\vec{v}) = -u^0 v^0 d\vec{x}_0 \wedge d\vec{x}_1 \wedge d\vec{x}_2 \wedge d\vec{x}_3 -$$
$$- u^1 v^1 d\vec{x}_1 \wedge d\vec{x}_0 \wedge d\vec{x}_2 \wedge d\vec{x}_3 +$$
$$+ u^2 v^2 d\vec{x}_2 \wedge d\vec{x}_0 \wedge d\vec{x}_1 \wedge d\vec{x}_3 -$$
$$- u^3 v^3 d\vec{x}_3 \wedge d\vec{x}_0 \wedge d\vec{x}_1 \wedge d\vec{x}_2$$
$$= \left(-u^0 v^0 + u^1 v^1 + u^2 v^2 + u^3 v^3\right) d\vec{x}_0 \wedge d\vec{x}_1 \wedge d\vec{x}_2 \wedge d\vec{x}_3$$
$$= \left(-u^0 v^0 + u^1 v^1 + u^2 v^2 + u^3 v^3\right) \vec{\omega}.$$

And finally

$$\begin{aligned}
\langle \vec{u}, \vec{v} \rangle &= (-1)^s * (\vec{u} \wedge (*\vec{v})) \\
&= (-1)^s * \left(\left(-u^0 v^0 + u^1 v^1 + u^2 v^2 + u^3 v^3 \right) \vec{\omega} \right) \\
&= (-1)^s \left(-u^0 v^0 + u^1 v^1 + u^2 v^2 + u^3 v^3 \right) * \vec{\omega} \\
&= (-1)^{2s} \left(-u^0 v^0 + u^1 v^1 + u^2 v^2 + u^3 v^3 \right) \\
&= -u^0 v^0 + u^1 v^1 + u^2 v^2 + u^3 v^3.
\end{aligned}$$

8.8   Dual map

A volume n-form $\tilde{\omega}$, Section 8.3, provides a mapping between a p-vector $\vec{v} \in \wedge^p \mathcal{V}$ and an $(n-p)$-form $\tilde{p} \in \wedge^{n-p} \mathcal{V}^\star$. This is called a *dual map*.

$$\tilde{\omega} : \wedge^p \mathcal{V} \longrightarrow \wedge^{n-p} \mathcal{V}^\star. \tag{8.35}$$

The form $\tilde{p}$ is called the dual of $\vec{v}$ with respect to $\tilde{\omega}$. It is constructed as follows. The number p of indices of the p-vector $v^{i_1 \dots i_p} = v^{[i_1 \dots i_p]}$ are contracted by the volume n-form $\omega_{k_1 \dots k_n} = \sqrt{|\det g|} \epsilon_{k_1 \dots k_n}$. The complementary remainder defines $p_{j_1 \dots j_{n-p}}$. And therefore

$$(n-p) - \text{form } \tilde{p} = \tilde{\omega}(p - \text{vector}), \tag{8.36}$$

or

$$p_{j_1 \dots j_{n-p}} = \frac{1}{p!} \, \omega_{i_1 \dots i_p j_1 \dots j_{n-p}} \, v^{i_1 \dots i_p}. \tag{8.37}$$

The factor $\frac{1}{p!}$ serves as a sort of normalization to avoid double counting due to the complete antisymmetry of $\omega_{i_1 \dots i_p j_1 \dots j_{n-p}}$ and $v^{i_1 \dots i_p}$. And from the complete antisymmetry of ω and $\vec{v}$ it concludes that $\tilde{p}$ is as well completely antisymmetric: $p_{j_1 \dots j_{n-p}} = p_{[j_1 \dots j_{n-p}]}$, an thus an $(n-p)$-form.

A metric tensor, Chapter 4, produces a one-to-one transformation from $(p,0)$-tensors to $(0,p)$-tensors, and vice versa.

$$\begin{aligned}
\mathbf{g} &: (p,0) - \text{tensor} \in \mathcal{V} \longrightarrow (0,p) - \text{tensor} \in \mathcal{V}^\star, \\
\mathbf{g} &: (0,p) - \text{tensor} \in \mathcal{V}^\star \longrightarrow (p,0) - \text{tensor} \in \mathcal{V}.
\end{aligned} \tag{8.38}$$

Thus, performing a metric transformation g after a dual map $\tilde{\omega}$ produces a total transformation from $\wedge^p \mathcal{V}$ to $\wedge^{n-p} \mathcal{V}$:

$$g \circ \tilde{\omega} : \wedge^p \mathcal{V} \xrightarrow{\tilde{\omega}} \wedge^{n-p} \mathcal{V}^\star \xrightarrow{g} \wedge^{n-p} \mathcal{V}. \tag{8.39}$$

It can be recognised that the successive execution of $\tilde{\omega}$ and g corresponds to the Hodge mapping $*$.

Example 8.6

And again we want to use the ordinary vector product from vector calculus to apply the dual map from equation (8.37). We will assume $\sqrt{|\det g|} = 1$. The dual map $\tilde{\omega}$ processes the indices of $(p, 0)$-tensors, and therefore we have to bring the vector product of two vectors $\vec{u}$ and $\vec{v}$ into a compact, antisymmetric component representation:

$$\vec{u} \times \vec{v} \equiv u^{[j} v^{k]}.$$

Inserted in (8.37)

$$p_l = \frac{1}{2!} \, \epsilon_{jkl} \, u^{[j} v^{k]} = \epsilon_{jkl} \, u^j v^k = \epsilon_{ljk} \, u^j v^k.$$

In the penultimate step, the correction factor for double counting was omitted because the vector components are no longer anti-symmetrical by removing the square bracket. In the last step, the indices of the Levi-Civita tensor were cyclically replaced. Finally, we perform a metric transformation and obtain the well-known formula for a vector product (see also Appendix D.1):

$$p_l = \epsilon_{ljk} \, u^j v^k,$$
$$g^{il} p_l = g^{il} \, \epsilon_{ljk} \, u^j v^k,$$
$$p^i = \epsilon^i{}_{jk} \, u^j v^k.$$

∎

Because of the known fact that the $(n - p)$-form and the p-vector from equation (8.36) have the same number of free components due to $C_p^n = \binom{n}{p} = \binom{n}{n-p}$, equation (8.3), the dual map $\tilde{\omega}(\vec{v})$, (8.36), is invertible. Thus, we define a volume n-vector

$$\tilde{\omega}^{-1} = \frac{1}{\sqrt{|\det g|}} \epsilon^{k_1 \ldots k_n} \tag{8.40}$$

that is inverse to the volume n-form:

$$\tilde{\omega}^{-1}\,\tilde{\omega} = \frac{1}{\sqrt{|\mathrm{det}g|}}\epsilon^{k_1\,\dots\,k_n}\,\sqrt{|\mathrm{det}g|}\epsilon_{k_1\,\dots\,k_n} = \epsilon^{k_1\,\dots\,k_n}\,\epsilon_{k_1\,\dots\,k_n} = n!. \qquad (8.41)$$

See in addition Appendix D.

A dual map of a p-form to an $(n-p)$-vector is thus defined as:

$$\tilde{\omega}^{-1} : \wedge^p \mathcal{V}^\star \longrightarrow \wedge^{n-p}\mathcal{V}. \qquad (8.42)$$

8.9 Exterior derivative

The *exterior derivative operator* $\tilde{\mathbf{d}}$ is an operator that is applied *exclusively to forms* including the 0-form (functions). The exterior derivative $\tilde{\mathbf{d}}$ is defined as an *anti-symmetrized partial derivative* of a form (completely antisymmetric tensor). So it transforms a p-form into a $(p+1)$-form:

$$\tilde{\mathbf{d}} : \wedge^p \mathcal{V}^\star \longrightarrow \wedge^{p+1} \mathcal{V}^\star. \tag{8.43}$$

We use the following notation for $\tilde{\mathbf{d}}$:

$$\tilde{\mathbf{d}} \equiv \frac{\partial}{\partial x^i} d\tilde{x}^i \wedge \equiv \partial_i d\tilde{x}^i \wedge . \tag{8.44}$$

In Chapter 5, "Gradient", Section 5.1 we learned about the exterior derivative operator $\tilde{\mathbf{d}}$. We used it there without the wedge symbol $\wedge$, because it was applied exclusively to scalar functions, and it is $\wedge f = f$. To complete it, let it be done again here, an exterior derivative on a 0-form f:

$$\tilde{\mathbf{d}} f = \left(\partial_i d\tilde{x}^i \wedge \right) f = \left(\partial_i f \right) d\tilde{x}^i \equiv \text{gradient of } f. \tag{8.45}$$

The application of the exterior derivative to a 1-form p is

$$\tilde{\mathbf{d}} \tilde{p} = \tilde{\mathbf{d}} \left(p_j d\tilde{x}^j \right) = \left(\partial_i p_j \right) d\tilde{x}^i \wedge d\tilde{x}^j. \tag{8.46}$$

For $n = 3$ the result is

$$\begin{aligned}
\left(\partial_i p_j \right) d\tilde{x}^i \wedge d\tilde{x}^j &= \left(\partial_1 p_2 \right) d\tilde{x}^1 \wedge d\tilde{x}^2 + \left(\partial_2 p_1 \right) d\tilde{x}^2 \wedge d\tilde{x}^1 + \\
&+ \left(\partial_1 p_3 \right) d\tilde{x}^1 \wedge d\tilde{x}^3 + \left(\partial_3 p_1 \right) d\tilde{x}^3 \wedge d\tilde{x}^1 + \\
&+ \left(\partial_2 p_3 \right) d\tilde{x}^2 \wedge d\tilde{x}^3 + \left(\partial_3 p_2 \right) d\tilde{x}^3 \wedge d\tilde{x}^2 \\
&= \left(\partial_1 p_2 - \partial_2 p_1 \right) d\tilde{x}^1 \wedge d\tilde{x}^2 + \\
&+ \left(\partial_1 p_3 - \partial_3 p_1 \right) d\tilde{x}^1 \wedge d\tilde{x}^3 + \\
&+ \left(\partial_2 p_3 - \partial_3 p_2 \right) d\tilde{x}^2 \wedge d\tilde{x}^3.
\end{aligned} \tag{8.47}$$

Here you can see the "similarity" to the *curl* operation from vector calculus. If one maps the expression one-to-one onto a 1-dimensional basis representation by means of the Hodge star operator $*$, the curl expression is obtained.

$$\begin{aligned}
* \left(\partial_i p_j \right) d\tilde{x}^i \wedge d\tilde{x}^j &= \left(\partial_2 p_3 - \partial_3 p_2 \right) d\tilde{x}^1 + \\
&+ \left(\partial_3 p_1 - \partial_1 p_3 \right) d\tilde{x}^2 + \\
&+ \left(\partial_1 p_2 - \partial_2 p_1 \right) d\tilde{x}^3.
\end{aligned} \tag{8.48}$$

In Euclidean space, the form components can be transferred 1:1 into vector components.

It is interesting to examine an exterior derivative on a volume form $\tilde{\omega}$ on an n-dimensional vector space $\wedge^n \mathcal{V}^\star$. Let the basis of the volume form be

$$\left\{ d\tilde{x}^I \right\} = \left\{ d\tilde{x}^{i_1} \wedge \ldots \wedge d\tilde{x}^{i_n} \right\}, \; i_1 < \ldots < i_n,$$

and thus

$$\tilde{\mathbf{d}}\,\tilde{\omega} = \tilde{\mathbf{d}}\,\omega_I d\tilde{x}^I = \left(\partial_i \omega_I \right) d\tilde{x}^i \wedge d\tilde{x}^I = 0, i \in I. \tag{8.49}$$

The result confirms the proposition that on an n-dimensional vector space the degree p of a form must satisfy the condition $p \leq n$, otherwise it vanishes.

Box 8.9

The exterior derivative operator $\tilde{\mathbf{d}}$ transforms like a tensor. We show this on a 1-form $\tilde{p}$.

$$\tilde{\mathbf{d}}\,\tilde{p} = \left(\partial_i p_j \right) d\tilde{x}^i \wedge d\tilde{x}^j,$$

$$\left(\tilde{\mathbf{d}}\,\tilde{p} \right)' = \left(\partial_{i'} p_{j'} \right) d\tilde{x}^{i'} \wedge d\tilde{x}^{j'}.$$

Let's transform $\partial_{i'} p_{j'}$

$$\frac{\partial}{\partial x^{i'}} p_{j'} = \frac{\partial x^i}{\partial x^{i'}} \frac{\partial}{\partial x^i} \left(\frac{\partial x^j}{\partial x^{j'}} p_j \right)$$

$$= \frac{\partial x^i}{\partial x^{i'}} \frac{\partial x^j}{\partial x^{j'}} \left(\frac{\partial}{\partial x^i} p_j \right) + p_j \frac{\partial^2 x^j}{\partial x^{i'} \partial x^{j'}}.$$

Since the 2-form components $\left(\partial_{i'} p_{j'} \right)$ is by definition a completely antisymmetric tensor, the symmetrical part $\left(p_j \frac{\partial^2 x^j}{\partial x^{i'} \partial x^{j'}} \right)$ (ordinary partial derivatives commutate) vanish. Refer to equation (8.2).

The exterior derivative operator $\tilde{\mathbf{d}}$ has the following properties where $\tilde{p}$ is a p-form and $\tilde{q}$, $\tilde{r}$ are q-forms:

$$
\begin{aligned}
(1)\quad &\tilde{\mathbf{d}}\left(\tilde{q}+\tilde{r}\right) &&= \tilde{\mathbf{d}}\,\tilde{q}+\tilde{\mathbf{d}}\,\tilde{r},\\
(2)\quad &\tilde{\mathbf{d}}\left(\tilde{p}\wedge\tilde{q}\right) &&= \left(\tilde{\mathbf{d}}\,\tilde{p}\right)\wedge\tilde{q}+(-1)^{p}\,\tilde{p}\wedge\left(\tilde{\mathbf{d}}\,\tilde{q}\right),\\
(3)\quad &\tilde{\mathbf{d}}\left(\tilde{\mathbf{d}}\,\tilde{p}\right) &&= 0,\\
(4)\quad &\tilde{\mathbf{d}}\,f &&= \partial_{i}f\,d\tilde{x}^{i},\ \text{for each smooth function } f.
\end{aligned}
\tag{8.50}
$$

The property (2) is according to the Leibniz rule with the additional factor $(-1)^{p}$, which is due to the fact that the operator $\tilde{\mathbf{d}}$ has to be shifted through the p-form $\tilde{p}$ to reach to $\tilde{q}$. Property (3) looks a little surprising, but is due to the fact that the partial derivative within the operator process $\tilde{\mathbf{d}}$ commutates, i.e. is symmetrical in the indices, whereas additionally introduced wedge product is anti-symmetrical. This property (3) is derived in detail below on a 0-form and on a 1-form. For a p-form see Box 8.10.

Let f be a 0-form, and then the following applies:

$$
\begin{aligned}
\tilde{\mathbf{d}}\left(\tilde{\mathbf{d}}\,f\right) &= \partial_{j}d\tilde{x}^{j}\wedge\left(\partial_{i}f d\tilde{x}^{i}\right) = \left(\partial_{j}\partial_{i}f\right)d\tilde{x}^{j}\wedge d\tilde{x}^{i}\\
&= -\left(\partial_{j}\partial_{i}f\right)d\tilde{x}^{i}\wedge d\tilde{x}^{j} \overset{\partial_{j}\partial_{i}=\partial_{i}\partial_{j}}{=} -\underbrace{\left(\partial_{i}\partial_{j}f\right)d\tilde{x}^{i}\wedge d\tilde{x}^{j}}_{\text{dummy } i\leftrightarrow j}\\
&= -\left(\partial_{j}\partial_{i}f\right)d\tilde{x}^{j}\wedge d\tilde{x}^{i},\\
&\Longrightarrow\\
&\left(\partial_{j}\partial_{i}f\right)d\tilde{x}^{j}\wedge d\tilde{x}^{i} = -\left(\partial_{j}\partial_{i}f\right)d\tilde{x}^{j}\wedge d\tilde{x}^{i},\\
&\Longrightarrow\\
\tilde{\mathbf{d}}\left(\tilde{\mathbf{d}}\,f\right) &= 0.
\end{aligned}
\tag{8.51}
$$

In the first line we have made use of the rule that the wedge operator $d\tilde{x}^{j}\wedge$ only processes basis elements but no scalars $\partial_{i}f$. In the second line, the ordinary partial derivatives commutated, and when the transition to the third line was made, the dummy indices were relabelled. Comparing the initial term $\tilde{\mathbf{d}}\left(\tilde{\mathbf{d}}\,f\right)$ with the executions concerning an exterior derivative of a 1-form, equation (8.47), one recognises in $\tilde{\mathbf{d}}\,f$ the 1-form *gradient* and in an exterior derivative of a 1-form the *curl* properties. In sum: curl grad $= 0$.

And the examination of a 1-form $\tilde{p} = p_k d\tilde{x}^k$ shows:

$$\tilde{\mathbf{d}}\left(\tilde{\mathbf{d}}\,\tilde{p}\right) = \tilde{\mathbf{d}}\left(\left(\partial_j d\tilde{x}^j \wedge\right) p_k d\tilde{x}^k\right) = \tilde{\mathbf{d}}\left(\left(\partial_j p_k\right) d\tilde{x}^j \wedge d\tilde{x}^k\right)$$

$$= \left(\partial_i \partial_j p_k\right) d\tilde{x}^i \wedge d\tilde{x}^j \wedge d\tilde{x}^k$$

$$= -\left(\partial_i \partial_j p_k\right) d\tilde{x}^j \wedge d\tilde{x}^i \wedge d\tilde{x}^k - \left(\partial_j \partial_i p_k\right) d\tilde{x}^j \wedge d\tilde{x}^i \wedge d\tilde{x}^k$$

$$= -\left(\partial_i \partial_j p_k\right) d\tilde{x}^i \wedge d\tilde{x}^j \wedge d\tilde{x}^k,$$

$$\Longrightarrow$$

$$\left(\partial_i \partial_j p_k\right) d\tilde{x}^i \wedge d\tilde{x}^j \wedge d\tilde{x}^k = -\left(\partial_i \partial_j p_k\right) d\tilde{x}^i \wedge d\tilde{x}^j \wedge d\tilde{x}^k,$$

$$\Longrightarrow$$

$$\tilde{\mathbf{d}}\left(\tilde{\mathbf{d}}\,\tilde{p}\right) = 0.$$

$$(8.52)$$

The mathematical operations are the same as in the example above: Swapping the dual basis that carry the index of the partial derivatives (change of sign), then commutating the partial derivatives (no change of sign), and finally relabelling the dummy indices. The result is an alleged identity of two expressions that differ only in sign. A geometric interpretation of what has just been examined: An exterior derivative of a 1-form (curl, see equation (8.47)), undergoes a second exterior derivative. In the result we see a volume form $\left(d\tilde{x}^i \wedge d\tilde{x}^j \wedge d\tilde{x}^k\right)$ for $n = 3$. In sum: divergence curl $= 0$.

Box 8.10

Let $\tilde{p}$ a p-form

$$\tilde{p} = p_I d\tilde{x}^I, \text{with } I = \{i_1 < \ldots < i_p, p \leq n\}.$$

And in the same procedure as in the examples above, we proceed with the p-form:

$$\tilde{\mathbf{d}}\left(\tilde{\mathbf{d}}\,\tilde{p}\right) = \tilde{\mathbf{d}}\left(\partial_j d\tilde{x}^j \wedge p_I d\tilde{x}^I\right) = \tilde{\mathbf{d}}\left(\partial_j p_I d\tilde{x}^j \wedge d\tilde{x}^I\right)$$

$$= \partial_k d\tilde{x}^k \wedge \left(\partial_j p_I d\tilde{x}^j \wedge d\tilde{x}^I\right)$$

$$= \partial_k \partial_j p_I d\tilde{x}^k \wedge d\tilde{x}^j \wedge d\tilde{x}^I$$

$$= -\partial_k \partial_j p_I d\tilde{x}^j \wedge d\tilde{x}^k \wedge d\tilde{x}^I$$

$$= -\partial_j \partial_k p_I d\tilde{x}^j \wedge d\tilde{x}^k \wedge d\tilde{x}^I$$

$$= -\partial_k \partial_j p_I d\tilde{x}^k \wedge d\tilde{x}^j \wedge d\tilde{x}^I,$$

$$\Longrightarrow$$

$$\partial_k \partial_j p_I d\tilde{x}^k \wedge d\tilde{x}^j \wedge d\tilde{x}^I = -\partial_k \partial_j p_I d\tilde{x}^k \wedge d\tilde{x}^j \wedge d\tilde{x}^I,$$

$$\Longrightarrow$$

$$\tilde{\mathbf{d}}\left(\tilde{\mathbf{d}}\,\tilde{p}\right) = 0.$$

From the transition from the third line to the fourth, the dual basis $d\tilde{x}^k$ and $d\tilde{x}^j$ were swapped, from the fourth line to the fifth, the partial derivatives were commutated and subsequently the dummy indices were relabelled.

Example 8.7

We want to express the following Maxwell's equation

$$\nabla \times \vec{E} + \frac{\partial}{\partial t}\vec{B} = 0,$$

with the vector fields $\vec{E}\left(E^1, E^2, E^3\right)$ and $\vec{B}\left(B^1, B^2, B^3\right)$, which generally depend on $\vec{r}$ and t, by using an exterior derivative and forms. We assign the vector field $\vec{E}$ to the 1-form $\tilde{\alpha}$ using the metric $\mathcal{V} \xrightarrow{\mathbf{g}} \mathcal{V}^\star$:

$$\tilde{\alpha} = E_1 d\tilde{x}^1 + E_2 d\tilde{x}^2 + E_3 d\tilde{x}^3.$$

And we assign the vector field $\vec{B}$ to a 2-form $\tilde{\beta}$ by using dual-mapping $\mathcal{V} \xrightarrow{\tilde{\omega}} \wedge^2 \mathcal{V}^\star$ (Section 8.8):

$$\tilde{\beta} = B_1 d\tilde{x}^2 \wedge d\tilde{x}^3 + B_2 d\tilde{x}^3 \wedge d\tilde{x}^1 + B_3 d\tilde{x}^1 \wedge d\tilde{x}^2.$$

By applying the exterior derivative to $\tilde{\alpha}$ we obtain

$$\mathbf{\tilde{d}}\,\tilde{\alpha} = (\partial_2 E_3 - \partial_3 E_2)\, d\tilde{x}^2 \wedge d\tilde{x}^3 +$$
$$+ (\partial_3 E_1 - \partial_1 E_3)\, d\tilde{x}^3 \wedge d\tilde{x}^1 +$$
$$+ (\partial_1 E_2 - \partial_2 E_1)\, d\tilde{x}^1 \wedge d\tilde{x}^2.$$

The Maxwell equation under review is in forms

$$\mathbf{\tilde{d}}\,\tilde{\alpha} + \frac{\partial \tilde{\beta}}{\partial t} = 0.$$

$\blacksquare$

8.10   Gradient, Curl and Divergence

This section serves as preparation for Section 8.11, which deals with Stoke's integration theorems. We show the equivalence of forms to a scalar expression, which in turn is suitable for integration. We will operate in Euclidean space $\mathbb{R}^3 : \partial_i \equiv \partial^i$, $d\tilde{x}^i \equiv d\tilde{x}_i$. This gives us comprehensible results that also apply to manifolds of arbitrary dimension and signature.

Gradient:
We start from a 0-form f and form the exterior derivative, see equation (8.45):

$$\mathbf{\tilde{d}}f = \left(\partial_i d\tilde{x}^i \wedge\right) f = \left(\partial_i f\right) d\tilde{x}^i.$$

With the conditions specified, we can state:

$$\mathbf{\tilde{d}}f \equiv \nabla f \cdot d\vec{r}. \tag{8.53}$$

Curl:

We start from a 1-form $\tilde{A}$ and form the exterior derivative, see equation (8.47):

$$\tilde{\mathbf{d}}\,\tilde{A} = \partial_i d\tilde{x}^i \wedge \left(A_1 d\tilde{x}^1 + A_2 d\tilde{x}^2 + A_3 d\tilde{x}^3\right)$$
$$= \left(\partial_2 A_3 - \partial_3 A_2\right) d\tilde{x}^2 \wedge d\tilde{x}^3 +$$
$$+ \left(\partial_3 A_1 - \partial_1 A_3\right) d\tilde{x}^3 \wedge d\tilde{x}^1 +$$
$$+ \left(\partial_1 A_2 - \partial_2 A_1\right) d\tilde{x}^1 \wedge d\tilde{x}^2.$$

With the conditions specified, we can state:

$$\tilde{\mathbf{d}}\,\tilde{A} \equiv \left(\nabla \times \vec{A}\right) \cdot d\vec{S}. \tag{8.54}$$

Divergence:

We start from a 2-form $\tilde{B}$ and form the exterior derivative

$$\tilde{\mathbf{d}}\,\tilde{B} = \partial_i d\tilde{x}^i \wedge \left(B_1 d\tilde{x}^2 \wedge d\tilde{x}^3 + B_2 d\tilde{x}^3 \wedge d\tilde{x}^1 + B_3 d\tilde{x}^1 \wedge d\tilde{x}^2\right)$$
$$= \left(\partial_1 B_1 + \partial_2 B_2 + \partial_3 B_3\right) d\tilde{x}^1 \wedge d\tilde{x}^2 \wedge d\tilde{x}^3.$$

With the conditions specified, we can state:

$$\tilde{\mathbf{d}}\,\tilde{B} = \left(\nabla \cdot \vec{B}\right) dV. \tag{8.55}$$

8.11   General Stoke's Theorem

The general Stoke's theorem can be summarized as follows:

$$
\underbrace{\int_{R}}_{\substack{R \\ \text{region } R}} \overset{\text{derivative}}{\tilde{d}} \;(\text{form}) = \int_{\substack{\partial R \\ \text{boundary of } R}} (\text{form}).
\qquad (8.56)
$$

where the left integral sign carries the label *anti-derivative*.

An integration of forms over an n-dimensional region is only defined by n-forms. Since the exterior derivative $\tilde{\mathbf{d}}$ and the integration $\int$ are inverse processes to each other, the following is given. If the n-form integrand is defined as in (8.56) with

$$
\tilde{\mathbf{d}} \underbrace{(\text{form})}_{n-1} = n - \text{form},
$$

then the $(n-1)$-form can only be integrated over an $(n-1)$-dimensional hypersurface, the *boundary* of the region.

The boundary of a globe is a sphere. The sphere itself has no boundary. The boundary of a disk is a circle. The circle itself has no boundary. The boundaries of a curve are its end points. A point itself has no boundary. Or in general:

$$
\partial \partial R = 0.
$$

This reminds us of the result from Section 8.9 that a form, regardless of its degree, which undergoes an exterior derivation performed twice in succession, vanishes:

$$
\tilde{\mathbf{d}}\,\tilde{\mathbf{d}}\,\tilde{p} = 0.
$$

And transferring these considerations equivalently to integration means:

$$
\int_{R} \tilde{\mathbf{d}}\left(\tilde{\mathbf{d}}\,\text{form}\right) = \int_{\partial R} \left(\tilde{\mathbf{d}}\,\text{form}\right) = \int_{\partial \partial R} (\text{form}) = 0.
\qquad (8.57)
$$

Together with the results from Section 8.10, we can deduce the following integrals, which are very familiar from physics and vector calculus, very easily.

Line integral

For a line integral the equation (8.56) requires a 0-form f and with (8.53) we get

$$\int_C \tilde{\mathbf{d}}\, f = \int_C \nabla f \cdot d\vec{r} = \int_{\partial C \equiv \mathcal{P}} f = f\Big|_A^B. \tag{8.58}$$

This is the *Fundamental Theorem of Line Integrals.*

Surface integral

For a surface integral the equation (8.56) requires a 1-form $\tilde{A}$ and with (8.54) we get

$$\int_S \tilde{\mathbf{d}}\, \tilde{A} = \int_S \left(\nabla \times \vec{A} \right) \cdot d\vec{S} = \oint_{\partial S \equiv C} \vec{A} \cdot d\vec{r}. \tag{8.59}$$

This is *Stoke's Theorem.*

Volume Integral

For a volume integral the equation (8.56) requires a 2-form $\tilde{B}$ and with (8.55) we get

$$\int_V \tilde{\mathbf{d}}\, \tilde{B} = \int_V \left(\nabla \cdot \vec{B} \right) dV = \int_{\partial V \equiv S} \vec{B} \cdot d\vec{S}. \tag{8.60}$$

This is the *Divergence Theorem* (Gauss's Law).

Chapter 9

Introduction to Special Relativity (SR)

9.1   Inertial frame

Reference frames (coordinates) that provide a linear and constant movement, are called *Galilean reference frames, inertial frames* or *inertial coordinates*. Identical physical experiments in different inertial frames lead to identical results. Therefore it is not possible to conclude from the result of an experiment whether it was carried out in a resting or uniformly moving coordinate system. Therefore the absolute speed of a coordinate system is not measurable for an "inhabitant" of this system. This is the well-known *principle of relativity*.

The three fundamental laws of Newtonian mechanics (NM) contain no velocities but only accelerations. Therefore NM is invariant to different inertial frames. Let $\vec{u}(t)$ be the velocity of a point (particle) relative to the inertial system $\mathcal{K}$. An inertial system $\mathcal{K}'$ moves relative to $\mathcal{K}$ with the constant velocity $\vec{v}$. Then, using the classical *Galilei transformation*, the velocity $\vec{u}'$ of this particle is relative to $\mathcal{K}'$

$$\vec{u}' = \vec{u}(t) - \vec{v}. \tag{9.1}$$

And the corresponding accelerations are calculated to:

$$\vec{a}'(t) = \frac{d\vec{u}'}{dt} = \frac{d\left(\vec{u}(t) - \vec{v}\right)}{dt}$$
$$= \frac{\vec{u}(t)}{dt} - \frac{\vec{v}(t)}{dt}$$
$$= \vec{a}(t) - 0 = \vec{a}(t).$$

From this it follows that in both inertial coordinates $\mathcal{K}$ and $\mathcal{K}'$ Newton's laws are identical, and thus *invariant* to transformations from one inertial frame to another.

The term *"special"* in the theory term 'special relativity' stems precisely from the fact that it refers *only* to inertial frames, and not to general coordinate systems, such as accelerated ones.

9.2   Minkowski space (MS)

Besides the postulate of invariance of physical processes between inertial frames (Section 9.1), there is a second one, the *constancy of the speed of light*: The speed of light $c = 3 \times 10^8 \, m/s$ of a light signal measured relative to one inertial frame $\mathcal{K}$ is identical with the measurement of this light signal relative to a second inertial frame $\mathcal{K}'$ moving at constant speed.

As a consequence, we have to extend the description of an inertial frame from Section 9.1 as follows: Each point $\mathcal{P}$ (spatial vector $\vec{r}$) of an inertial coordinate system is equipped with a clock, time component t, which runs synchronously to each other, the so-called *proper time*. The coordinate system of SR thus consists of coordinate points $(t, \vec{r})$, which are called *events*. An event is therefore defined by a location and a time information to an inertial frame $\mathcal{K}(t, \vec{r})$. A four-dimensional flat space of events $\left(t, \vec{r}\left(x^1, x^2, x^3\right)\right)$ in an orthogonal coordinate system is called *Minkowski*[1] *Space*. When we speak of the Minkowski space in the present text, we use the short form MS or in modern terms called *spacetime*, shortened ST.

Remark 9.1
At the beginning of this chapter we speak mainly of *events* as coordinate points to emphasize the indissoluble connection of time and space in the Minkowski space of the SR. Only later in the chapter will we return to the usual term *'point'*.

■

---

[1]Hermann Minkowski, 1864-1909

The distance $(\triangle s)^2$ or *interval* between two events of an inertial frame $\mathcal{K}$ is defined as follows:

$$(\triangle s)^2 = -c^2 \left(\triangle t\right)^2 + \left(\triangle \vec{r}\right)^2 . \tag{9.2}$$

Definition 9.1
In order not always to carry the "ballast" of the large number $c = 3 \times 10^8 \, m/s$ in the coordinates, the speed of light is set to 1, $c = 1$. Thus the time coordinate t gets the dimension of a length. We are already accustomed to this in astronomy, where distances from stars are given in time: the light years.

$$c = 3 \times 10^8 \, ms^{-1} = 1,$$
$$1\,s = 3 \times 10^8 \, m,$$
$$1\,m = \frac{1}{3 \times 10^8} \, s.$$

If in the further course of the text we give information on a velocity v, we speak of a dimensionless quantity $v \left[\frac{m}{s} = \frac{m}{s=3\times 10^8} = \frac{1}{3\times 10^8} \right]$, which represents the proportion of the speed of light.

$\blacksquare$

And so we adjust our equation for the interval of two events equation (9.2) with $c = 1$:

$$(\triangle s)^2 = -\left(\triangle t\right)^2 + \left(\triangle \vec{r}\right)^2 . \tag{9.3}$$

The physical measurement of a light signal emitted at $(t_1, \vec{r}_1)$ and received at $(t_2, \vec{r}_2)$ follows the condition

$$|\vec{r}_2 - \vec{r}_1| = t_2 - t_1,$$
$$(\triangle \vec{r})^2 = (\triangle t)^2 , \tag{9.4}$$
$$0 = -\left(\triangle t\right)^2 + \left(\triangle \vec{r}\right)^2 .$$

The result is remarkable: light signal events have a zero distance! These events are called *lightlike* or *null separated*. Since a light signal always propagates at the same speed, regardless of the inertial frame $\mathcal{K}$ or $\mathcal{K}'$, the same applies for $\mathcal{K}'$:

$$0 = -(\triangle t')^2 + (\triangle \vec{r}')^2. \tag{9.5}$$

It can be shown (see Box 9.1) that in general

$$(\triangle s)^2 = (\triangle s')^2, \text{ or } ds^2 = ds'^2. \tag{9.6}$$

The difference value between two events, the spacetime interval, the distance between two points, is independent of the selected inertial frame and thus an invariant quantity. One changes the inertial coordinates without influencing the event distances; i.e. these intervals are a *property of the physical process,* but not of the coordinate system. *This invariance of the interval is a fundamental theorem of the SR.* The events with $ds^2 < 0$ are called *timelike separated,* while the events with $ds^2 > 0$ are called *spacelike separated.*

Box 9.1   invariance of the interval, equation (9.6)
The following explanation follows the reasoning of L.D.Landau, E.M.Lifschitz "Theoretische Physik", Vol.2.

Since the infinitesimal intervals ds^2 and ds'^2 are scalars and of the same order, the transformation $ds^2 = F\, ds'^2$ can only consist of a dimensionless transformation coefficient k: $ds^2 = k\, ds'^2$, which depends only on the relation between the inertial frames $\mathcal{K}$ and $\mathcal{K}'$. $\mathcal{K}$ and $\mathcal{K}'$ differ only in the relative velocity $\vec{v}$, (dimensionless, see Definition 9.1), they have to each other. Here the magnitude of $|\vec{v}|$ must be considered, because a directional dependence would violate the isotropy of space.

We consider three inertial frames $\mathcal{K}1, \mathcal{K}2$ and $\mathcal{K}3$. The relative velocity between $\mathcal{K}1$ and $\mathcal{K}2$ is v_{21}, and that between $\mathcal{K}2$ and $\mathcal{K}3$ is v_{32}. Then obviously the following relationship must apply between the transformation coefficients k

$$\underbrace{k\,(v_{31})}_{\mathcal{K}1\longrightarrow\mathcal{K}3} = \underbrace{k\,(v_{32})}_{\mathcal{K}2\longrightarrow\mathcal{K}3}\ \underbrace{k\,(v_{21})}_{\mathcal{K}1\longrightarrow\mathcal{K}2},$$

where v_{31} is the relative velocity between $\mathcal{K}1$ and $\mathcal{K}3$. The LHS depends on the angle that the speeds v_{21} and v_{32} include. The RHS has no reference to an angle. This contradiction is only resolved by the property that k is constant. And thus: $k = k^2$. The only non-trivial, real solution is: $k = 1$.

This proved the validity of the relationship from (9.6)

The canonical metric tensor of the MS is

$$\eta = \begin{pmatrix} -1 & 0 & 0 & 0 \\ 0 & 1 & 0 & 0 \\ 0 & 0 & 1 & 0 \\ 0 & 0 & 0 & 1 \end{pmatrix}, \quad \eta_{ij} = \begin{cases} -1 & \text{for } i = j = 0 \\ 1 & \text{for } i = j = 1, 2, 3 \\ 0 & \text{for } i \neq j \end{cases} \tag{9.7}$$

The indices start with the number 0, representing the time component. The space components run as usual in geometry and vector calculus in ascending order over the natural numbers. And so we can express the line element ds^2 from (9.6) using the Minkowski metric as follows:

$$\begin{aligned} ds^2 &= \eta_{ij} dx^i dx^j \\ &= \eta_{00} \left(dx^0\right)^2 + \eta_{11} \left(dx^1\right)^2 + \eta_{22} \left(dx^2\right)^2 + \eta_{33} \left(dx^3\right)^2 \\ &= -dt^2 + \left(dx^1\right)^2 + \left(dx^2\right)^2 + \left(dx^3\right)^2, \\ &= -dt^2 + dx^2 + dy^2 + dz^2. \end{aligned} \tag{9.8}$$

Since ds^2 is a *frame invariant scalar quantity*, (9.6), it is called *proper time* for a time like event, $ds^2 < 0$,

$$d\tau^2 = -\eta_{ij} dx^i dx^j, \tag{9.9}$$

and *proper length* for a *space like* event, $ds^2 > 0$,

$$ds^2 = \eta_{ij} dx^i dx^j. \tag{9.10}$$

Remark 9.2

The proper time is not a constant, as the first impression from the name might suggest. Each point $\mathcal{P}(x^i)$ of the manifold MS has its own proper time $d\tau = \sqrt{-\eta_{ij} dx^i dx^j}$. $d\tau$ is a scalar and thus invariant to Lorentz transformations (see Box 9.3).

■

Based on equation (9.8), the question can be asked at which geometric locations are the event intervals ds^2 at equal distances from the origin of the inertial coordinates $\mathcal{K}$? For the geometrical answer we restrict ourselves to a 2-dimensional MS with $\eta = \begin{pmatrix} -1 & 0 \\ 0 & 1 \end{pmatrix}$, i.e. with the coordinates (t, x). Since we refer to the origin: $dt = t$, $dx = x$. We thus obtain three equations:

$$-t^2 + x^2 = -d_1{}^2 < 0, \text{ timelike,}$$
$$-t^2 + x^2 = 0, \text{ lightlike or null separated,} \qquad (9.11)$$
$$-t^2 + x^2 = d_2{}^2 > 0, \text{ spacelike.}$$

The geometric location for constant timelike and spacelike intervals yields *hyperbolas*. Lightlike intervals are located on the straight lines $t = \pm x$. We enter these three position curves into a (t, x)-coordinate system and obtain the following Figure 9.1. In this diagram the lightlike events form a kind of

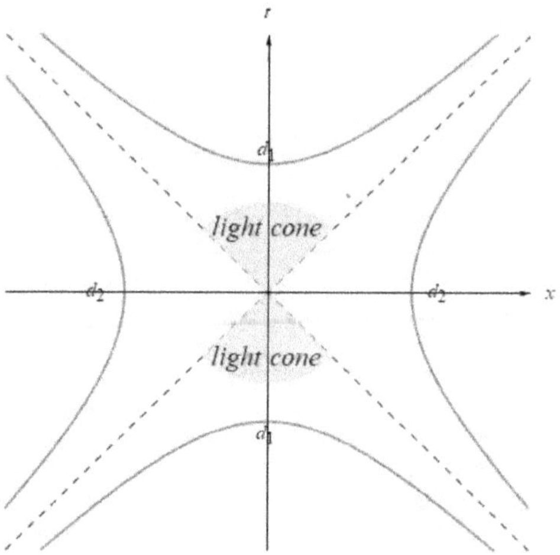

Figure 9.1: Invariant interval hyperbolas for the distances d_1 (timelike) and d_2 (spacelike). The dashed straight lines, asymptotes of the hyperbolas, denote the lightlike events.

dividing line between the timelike and spacelike events. If one were to add the y-coordinate in Figure 9.1, one would obtain a cone surface, the so-called *light cone*, in place of the two lightlike straight lines. Since physical objects can only move at speeds below the speed of light, physical events are therefore only timelike events, and are therefore only located inside the light cone. The path of a particle in spacetime, i.e. the *sequence of events*, is called a world line.

The *spacetime diagram* in Figure 9.1 has a different geometric meaning than what we are used to from geometric drawings. The points of the shown hyperbolas have an equal distance from the origin, although they seem to move further and further away from the origin the further one walks along the hyperbola branch. It is important to make this mental change, because spacetime diagrams are a powerful tool to describe processes in MS. We will discuss this in more detail in Section 9.4. The fact that time is assigned its own coordinate axis is due to the clearer graphic representation in spacetime diagrams. Actually, *all points are inherently assigned their own clock*, which run synchronously with each other. We want to represent the hyperbolic branch d_1 from Figure 9.1 in the way just described, that each point has an *inherent clock* and the same event is observed from different points (Figure 9.2).

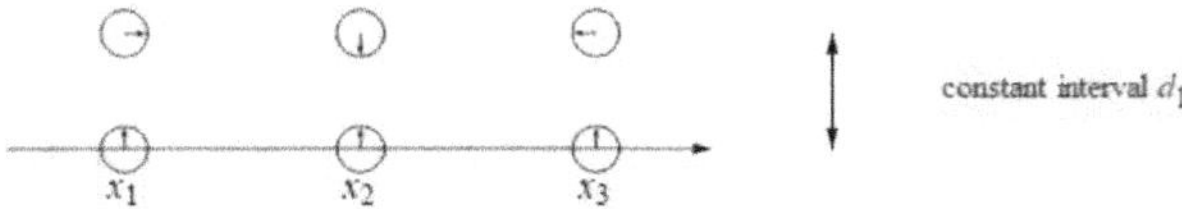

Figure 9.2: Hyperbola d_1 from Figure 9.1 in an alternative representation of clocks attached to positions.

In a two-dimensional Euclidean space with the metric $g_{ij} = \begin{pmatrix} 1 & 0 \\ 0 & 1 \end{pmatrix}$ we would obtain a *circle* for the location of the points which are at the same distance from the origin. In MS (2-dimensional), *hyperbolas* indicate equal distances (see Figure 9.3).

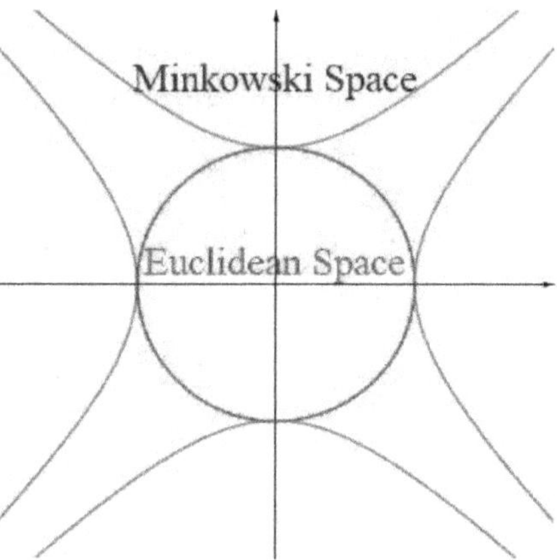

Figure 9.3: Pictorial representation of constant intervals of magnitude 1 for the Euclidean space and the Minkowski space of special relativity..

Box 9.2   *philosophical perspective*

Figure 9.3 shows the change in world view in the 19th century. Until the 19th century, the 'Weltanschauung' of mankind was a closed one. It was finite and "traceable". When physics gained new insights into the micro- and macrocosm at the transition to the 20th century, the world view changed dramatically. A "closed circle" became an "open hyperbole". Man was thrown out of his cosy warm nest into the infinitely and cold universe.

Summary 9.1

The *special theory of relativity*, founded by Albert Einstein (1879-1955), is based on three principles:

1. The *principle of relativity*: Identical experiments in different inertial frames lead to identical results. The name *special* relativity expresses the restriction of the reference frames to the *inertial* frames.

2. *Universality of the speed of light*: The speed of light $c = 3 \times 10^8$ m/s of a light signal measured relative to one inertial frame $\mathcal{K}$ is identical with the measurement of this light signal relative to a second inertial frame $\mathcal{K}'$.

3. *Transformation invariance*: The interval of any two events in a four-dimensional flat spacetime is independent of the inertial frame in which it is measured.

9.3 Lorentz transformation (LT)

There are three different types of coordinate transformations that leave intervals invariant in spacetime, equation 9.6:
1. *Translations* by simply shifting space and time coordinates (4 free parameters by one time coordinate and by 3 space coordinates).
2. *Rotations* around the space axes (3 free parameters by 3 space coordinates).
3. Relocations through a constant velocity vector, so-called *boosts* (3 free parameters through 3 space coordinates).
Rotations and boosts belong to the group of *Lorentz*[2] *transformations* (LT). LT together with the translations form the *Poincaré*[3] *group*.

We now want to derive the transformation matrix of a boost LT. We make the following assumptions: We consider the spacetime intervals between $(t, \vec{x})$ and $(0,0)$ in $\mathcal{K}$, $(\triangle s)^2 = -t^2 + \vec{x}^2$, and between $(t', \vec{x}')$ and $(0,0)$ in $\mathcal{K}'$, $(\triangle s')^2 = -t'^2 + \vec{x}'^2$. $(0,0)$ defines the origin in space and time respectively for the frames $\mathcal{K}$ and $\mathcal{K}'$. Without restriction of the generality we assume parallel

---

[2] Hendrik Antoon Lorentz, 1853-1928
[3] Henri Poincaré, 1854-1912

coordinates of the orthogonal coordinate systems of $\mathcal{K}$ and $\mathcal{K}'$ with coincident y and z axes. What is sought is a transformation

$$t = u(t',x'), \ x = v(t',x'),\tag{9.12}$$

which fulfils (9.6)

$$(\triangle s')^2 = (\triangle s)^2,$$
$$-(\triangle t')^2 + (\triangle x')^2 = -(\triangle t)^2 + (\triangle x)^2\tag{9.13}$$
$$= -u(t',x')^2 + v(t',x')^2$$

for all $(\triangle t, \triangle x)$ and $(\triangle t', \triangle x')$. This is only possible if u and v are *linear functions*. We do the transformation approach:

$$t = At' + Bx',$$
$$x = Ct' + dx',$$
$$y = y',$$
$$z = z',\tag{9.14}$$

$$\begin{pmatrix} t \\ x \\ y \\ z \end{pmatrix} = \begin{pmatrix} A & B & 0 & 0 \\ C & D & 0 & 0 \\ 0 & 0 & 1 & 0 \\ 0 & 0 & 0 & 1 \end{pmatrix} \begin{pmatrix} t' \\ x' \\ y' \\ z' \end{pmatrix}.$$

And thus we obtain the following relationship for an interval invariance, (9.13):

$$-t'^2 + x'^2 = -t^2 + x^2$$
$$= -\left(At' + Bx'\right)^2 + \left(Ct' + dx'\right)^2\tag{9.15}$$
$$= -(A^2 - C^2)t'^2 + (D^2 - B^2)x'^2 + 2(CD - AB)t'x'.$$

For uniqueness, the following must apply

$$A^2 - C^2 = 1, \ D^2 - B^2 = 1, \ CD - AB = 0.\tag{9.16}$$

The parametrisation of the coefficients A, B, C and D

$$A = D = \cosh\phi, \ B = C = \sinh\phi$$

results in the following solution for the transformation

$$t = t' \cosh \phi + x' \sinh \phi,$$

$$x = t' \sinh \phi + x' \cosh \phi,$$

$$y = y',$$

$$z = z',$$

$$\begin{pmatrix} t \\ x \\ y \\ z \end{pmatrix} = \begin{pmatrix} \cosh \phi & \sinh \phi & 0 & 0 \\ \sinh \phi & \cosh \phi & 0 & 0 \\ 0 & 0 & 1 & 0 \\ 0 & 0 & 0 & 1 \end{pmatrix} \begin{pmatrix} t' \\ x' \\ y' \\ z' \end{pmatrix}. \tag{9.17}$$

So far we have only dealt with the Minkowski metric for an interval invariance between two frames $\mathcal{K}$ and $\mathcal{K}'$. Now we want to introduce the *boost* condition and thus determine the physical meaning of the parameter ϕ. The boost condition is as follows: $\mathcal{K}'$ moves with the relative speed v along the x-,x'-coordinate in *positive direction*. We determine the physical significance of the parameter ϕ by considering the origin $x' = 0$ of $\mathcal{K}'$. The point $x' = 0$ moves, with the coordinate orientation between $\mathcal{K}$ and $\mathcal{K}'$ described above, at the speed $v = \frac{x}{t}$ along the x-coordinate. $x' = 0$ inserted in equation 9.17 yields

$$t = t' \cosh \phi, \ x = t' \sinh \phi, \ \Longrightarrow \ \frac{x}{t} = v = \tanh \phi. \tag{9.18}$$

This shows that it is equally true in relativistic mechanics that two inertial frames can only move relative to each other with a *constant* relative velocity $\tanh \phi$.

For hyperbolic functions the following conversions

$$\cosh \phi = \frac{1}{\sqrt{1 - \tanh^2 \phi}} = \frac{1}{\sqrt{1 - v^2}},$$

$$\sinh \phi = \frac{\tanh \phi}{\sqrt{1 - \tanh^2 \phi}} = \frac{v}{\sqrt{1 - v^2}} \tag{9.19}$$

apply, and we obtain the transformation between two inertial frames with the above mentioned conditions, the so-called *Lorentz-Transformation*. It is also called a *boost*, a displacement by means of a constant velocity vector. The Lorentz transformation from $\mathcal{K}'$ to $\mathcal{K}$, i.e. the (t', x') coordinates of $\mathcal{K}'$ are related to the (t, x) coordinates of $\mathcal{K}$, is given by

$$x^i = L^i_{j'} x^{j'}, \; i, j' = 0, 1, 2, 3, \text{ with } x^0, x^{0'} = t, t',$$

$$L^i_{j'} = \begin{pmatrix} \cosh\phi & \sinh\phi & 0 & 0 \\ \sinh\phi & \cosh\phi & 0 & 0 \\ 0 & 0 & 1 & 0 \\ 0 & 0 & 0 & 1 \end{pmatrix},$$

$$= \begin{pmatrix} \gamma & \gamma v & 0 & 0 \\ \gamma v & \gamma & 0 & 0 \\ 0 & 0 & 1 & 0 \\ 0 & 0 & 0 & 1 \end{pmatrix}, \text{ with } \gamma = \frac{1}{\sqrt{1 - v^2}},$$

$$t = t' \cosh\phi + x' \sinh\phi,$$

$$x = t' \sinh\phi + x' \cosh\phi,$$

$$y = y', z = z',$$

$$t = \gamma(t' + v x'),$$

$$x = \gamma(v t' + x'),$$

$$y = y', z = z'.$$

$$(9.20)$$

The representation from $\mathcal{K}$ to $\mathcal{K}'$ is also often chosen:

$$x^{i'} = L^{i'}_j x^j, \; i', j = 0, 1, 2, 3, \text{ with } x^0, x^{0'} = t, t',$$

$$L^{i'}_j = \begin{pmatrix} \cosh\phi & -\sinh\phi & 0 & 0 \\ -\sinh\phi & \cosh\phi & 0 & 0 \\ 0 & 0 & 1 & 0 \\ 0 & 0 & 0 & 1 \end{pmatrix},$$

$$= \begin{pmatrix} \gamma & -\gamma v & 0 & 0 \\ -\gamma v & \gamma & 0 & 0 \\ 0 & 0 & 1 & 0 \\ 0 & 0 & 0 & 1 \end{pmatrix}, \text{ with } \gamma = \frac{1}{\sqrt{1 - v^2}},$$

$$t' = t \cosh\phi - x \sinh\phi,$$

$$x' = -t \sinh\phi + x \cosh\phi,$$

$$y' = y, z' = z,$$

$$t' = \gamma(t - v x),$$

$$x' = \gamma(-v t + x),$$

$$y' = y, z' = z.$$

$$(9.21)$$

Definition 9.2

In order not to have to describe the coordinate orientations and motion relations between the frames $\mathcal{K}$ and $\mathcal{K}'$ always from new, we define the following short formula: An x^+-boost is based on the conditions according to (9.20), $\mathcal{K}' \longrightarrow \mathcal{K}$, or to (9.21), $\mathcal{K} \longrightarrow \mathcal{K}'$. In the case of an x^--boost, the frame $\mathcal{K}'$ moves with the speed v in the negative x-direction.

$\blacksquare$

To return to the beginning of the chapter on the Galilean transformation, let us examine the question of whether the LT contains the Galilean transformation as a limiting case. To carry out this investigation with equation (9.21), it is necessary to adapt this formula to the classical dimensions of time $[s]$, length $[m]$ and speed $[m/s]$:

$$t' = \gamma(t - \frac{v}{c^2} x),$$
$$x' = \gamma(-v\,t + x).$$

For $v \ll c : \gamma \longrightarrow 1, \frac{v}{c^2} \longrightarrow 0$, we get the classical formula of the Galilei transformation, (9.1),

$$t' = t,$$
$$x' = x - vt,$$
$$\Longrightarrow$$
$$\frac{dx'}{dt} = \frac{dx}{dt} - v.$$

To investigate the relativistic addition of velocities under the condition of the Lorentz transformation, let us make the following examination. We consider any object in $\mathcal{K}'$ which moves in x'-direction with the speed $u^{x'}$. We take the differentials of the expressions in equation (9.20)

$$dt = \gamma(dt' + vdx'),$$
$$dx = \gamma(vdt' + dx'),$$
$$dy = dy', \quad dz = dz',$$

and if quotients are calculated, the result is as follows

$$\frac{dx}{dt} = \frac{\gamma(vdt' + dx')}{\gamma(dt' + vdx')} = \frac{v + \frac{dx'}{dt'}}{1 + v\frac{dx'}{dt'}},$$

$$u^x = \frac{v + u^{x'}}{1 + vu^{x'}}, \tag{9.22}$$

$$\frac{dy}{dt} = \frac{dy'}{\gamma(dt' + vdx')} = \frac{\frac{dy'}{dt'}}{\gamma(1 + v\frac{dx'}{dt'})} = \frac{0}{\gamma(1 + v\frac{dx'}{dt'})} = 0,$$

$$\frac{dz}{dt} = 0.$$

Where u^x is the velocity of the object moving in x-direction relative to $\mathcal{K}$, and $u^{x'}$ is the velocity of this object relative to $\mathcal{K}'$. Here the question arises whether with the above velocity addition formula (9.22), a velocity greater than the speed of light can be observed? We assume that $u^{x'}$ and v are each less than 1, thus showing that u^x always remains less than 1 and is therefore below the speed of light.

$$u^{x'} - v\,u^{x'} = u^{x'}(1 - v) < 1 - v,$$

$$u^{x'} + v < 1 + v\,u^{x'},$$

$$\frac{u^{x'} + v}{1 + v\,u^{x'}} < 1,$$

$$\Longrightarrow$$

$$u^x = \frac{u^{x'} + v}{1 + v\,u^{x'}} < 1. \tag{9.23}$$

Box 9.3

spacetime metric tensor η and Lorentz transformation matrix L

The validity of postulate 3 of SR from the Summary 9.1, $ds'^2 = ds^2$, we also want to verify on the basis of the metric tensor η of spacetime, (9.7). With the Lorentz transformations matrix L (9.21) we can state that

$$ds'^2 = \eta_{i'j'}\, dx^{i'} dx^{j'} = \eta_{i'j'} \left(L_i^{i'}\, dx^i \right) \left(L_j^{j'}\, dx^j \right)$$

$$= \left(L_i^{i'} L_j^{j'}\, \eta_{i'j'} \right) dx^i dx^j$$

$$= \eta_{ij} dx^i dx^j = ds^2.$$

The transition from the second to the third row, contraction of the indices i', j', we want to reproduce explicitly with the matrices L and η. We adopt the reasoning of Section 4.3, (4.11), (4.12),

$$L_i^{i'} L_j^{j'}\, \eta_{i'j'} \quad \overset{\text{matrix form}}{\Longrightarrow} \quad \left(L_i^{i'} \right)^T \left(\eta_{j'}^{i'} \right) \left(L_j^{j'} \right)$$

and thus obtain

$$\left(L_i^{i'} \right)^T \left(\eta_{j'}^{i'} \right) \left(L_j^{j'} \right) =$$

$$= \begin{pmatrix} \gamma & -\gamma v & 0 & 0 \\ -\gamma v & \gamma & 0 & 0 \\ 0 & 0 & 1 & 0 \\ 0 & 0 & 0 & 1 \end{pmatrix}^T \begin{pmatrix} -1 & 0 & 0 & 0 \\ 0 & 1 & 0 & 0 \\ 0 & 0 & 1 & 0 \\ 0 & 0 & 0 & 1 \end{pmatrix} \begin{pmatrix} \gamma & -\gamma v & 0 & 0 \\ -\gamma v & \gamma & 0 & 0 \\ 0 & 0 & 1 & 0 \\ 0 & 0 & 0 & 1 \end{pmatrix}$$

$$= \begin{pmatrix} -1 & 0 & 0 & 0 \\ 0 & 1 & 0 & 0 \\ 0 & 0 & 1 & 0 \\ 0 & 0 & 0 & 1 \end{pmatrix} = \left(\eta_j^i \right).$$

The result is that the spacetime metric tensor η is invariant under the Lorentz transformation, and this also proves that the interval or distance in MS is invariant under the Lorentz transformation.

All transformations that keep the MS metric tensor η invariant, i.e.

$$\eta = L^T \eta L, \tag{9.24}$$

form a group of matrix multiplications, the so-called *Lorentz group*. This also includes, compare the first paragraph of this section, for example, rotations of the $x - y$ plane:

$$L = L_i^{i'} = \begin{pmatrix} 1 & 0 & 0 & 0 \\ 0 & \cos\theta & \sin\theta & 0 \\ 0 & -\sin\theta & \cos\theta & 0 \\ 0 & 0 & 0 & 1 \end{pmatrix},$$

$$L^T \eta L = \begin{pmatrix} 1 & 0 & 0 & 0 \\ 0 & \cos\theta & \sin\theta & 0 \\ 0 & -\sin\theta & \cos\theta & 0 \\ 0 & 0 & 0 & 1 \end{pmatrix}^T \begin{pmatrix} -1 & 0 & 0 & 0 \\ 0 & 1 & 0 & 0 \\ 0 & 0 & 1 & 0 \\ 0 & 0 & 0 & 1 \end{pmatrix} \begin{pmatrix} 1 & 0 & 0 & 0 \\ 0 & \cos\theta & \sin\theta & 0 \\ 0 & -\sin\theta & \cos\theta & 0 \\ 0 & 0 & 0 & 1 \end{pmatrix}$$

$$= \begin{pmatrix} -1 & 0 & 0 & 0 \\ 0 & 1 & 0 & 0 \\ 0 & 0 & 1 & 0 \\ 0 & 0 & 0 & 1 \end{pmatrix} = \eta.$$

9.4   Spacetime diagram

A *spacetime diagram* of $\mathcal{K}$ is an illustrative representation of a x^+-boost Lorentz transformation in the $x - t$ plane of $\mathcal{K}$. The relative velocity between $\mathcal{K}$ and $\mathcal{K}'$ is $v = \tanh\phi$, equation (9.18). The x'-coordinate of $\mathcal{K}'$ is determined by $t' = 0$ and therefore holds with equation (9.21)

$$\begin{aligned} 0 &= t \cosh\phi - x \sinh\phi, \\ t &= x \tanh\phi, \end{aligned} \tag{9.25}$$

and the t'-coordinate of $\mathcal{K}'$ is determined by $x' = 0$ and thus

$$\begin{aligned} 0 &= -t \sinh\phi + x \cosh\phi, \\ x &= t \tanh\phi. \end{aligned} \tag{9.26}$$

The $\mathcal{K}$ spacetime diagram in Figure 9.4 relates the (t', x') coordinates of the inertial frame $\mathcal{K}'$ to the (t, x) coordinates of the inertial frame $\mathcal{K}$ derived by means of Lorentz transformation. The representation of the $\mathcal{K}'$ coordinates in the $\mathcal{K}$ frame, lead to a collapse of the (t', x') coordinates, although they naturally remain orthogonal in the $\mathcal{K}'$ frame. Example 9.1 shows the $\mathcal{K}'$ space-

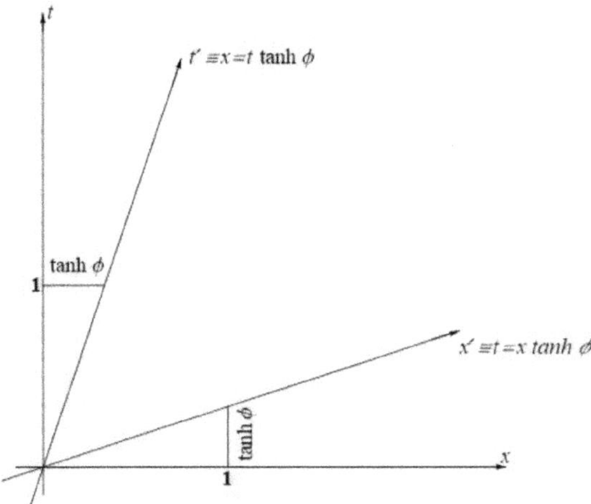

Figure 9.4: $\mathcal{K}$ spacetime diagram: The (t', x') coordinates of $\mathcal{K}'$ related to the (t, x) coordinates of $\mathcal{K}$ in the case of a x^+-boost Lorentz transformation.

time diagram of an x^+-boost, i.e. the (t, x) coordinates of $\mathcal{K}$ are related to the (t', x') coordinates of $\mathcal{K}'$.

Example 9.1

(t, x) coordinates of $\mathcal{K}$ related to the (t', x') coordinates of $\mathcal{K}'$ in the case of a x^+-boost Lorentz transformation.

Figure 9.5 shows the spacetime diagram of $\mathcal{K}'$ according to equation (9.20), the (t, x) coordinates in relation to the (t', x') coordinates. The x coordinate of $\mathcal{K}$ is determined by $t = 0$ and thus

$$0 = t' \cosh \phi + x' \sinh \phi,$$
$$t' = -x' \tanh \phi,$$

and the t coordinate of $\mathcal{K}$ is determined by $x = 0$ and thus

$$0 = t' \sinh \phi + x' \cosh \phi,$$
$$x' = -t' \tanh \phi.$$

The representation of the $\mathcal{K}$ coordinates in the $\mathcal{K}'$ frame, lead to an unfolding of the (t, x) coordinates, although they naturally remain orthogonal in the $\mathcal{K}$ frame.

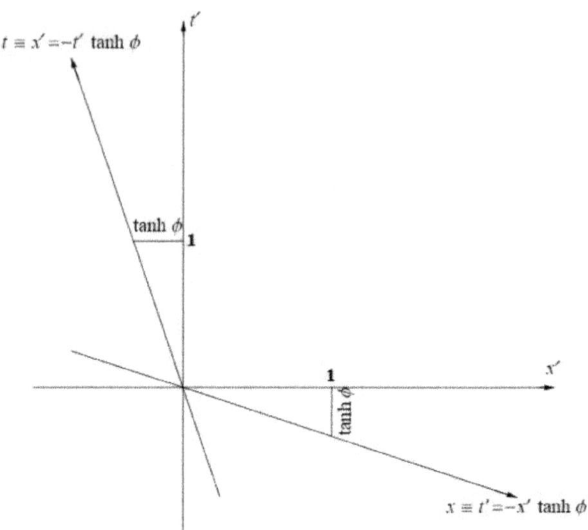

Figure 9.5: $\mathcal{K}'$ spacetime diagram: The (t, x) coordinates of $\mathcal{K}$ related to the (t', x') coordinates of $\mathcal{K}'$ in the case of a x^+-boost Lorentz transformation.

All further considerations, in this section and next sections, which we carry out for the spacetime diagram of $\mathcal{K}$, are of course to be applied in the same way for a spacetime diagram of $\mathcal{K}'$.

We want to perform a Lorentz transformation corresponding to equation (9.21) not only by calculation, but also graphically with the use of a spacetime diagram as shown in Figure 9.4. We consider the timelike event $\mathcal{A}$ with the coordinates $(2, 1)$ relative to frame $\mathcal{K}$, and calculate the coordinates (t', x') of this event $\mathcal{A}$ relative to frame $\mathcal{K}'$ with (9.21). The inertial frame $\mathcal{K}'$ moves with velocity $v = 1/3 = \tanh \phi$ relative to the inertial frame $\mathcal{K}$ in the positive

x-direction. And so we obtain:

$$t' = t \cosh \phi - x \sinh \phi$$

$$= 2 \cdot \frac{1}{\sqrt{1 - \tanh^2 \phi}} - 1 \cdot \frac{\tanh \phi}{\sqrt{1 - \tanh^2 \phi}} = \frac{2}{\sqrt{1 - (\frac{1}{3})^2}} - \frac{\frac{1}{3}}{\sqrt{1 - (\frac{1}{3})^2}} = 1.768,$$

$$x' = -t \sinh \phi + x \cosh \phi$$

$$= -2 \cdot \frac{\tanh \phi}{\sqrt{1 - \tanh^2 \phi}} + 1 \cdot \frac{1}{\sqrt{1 - \tanh^2 \phi}} = \frac{-\frac{2}{3}}{\sqrt{1 - (\frac{1}{3})^2}} + \frac{1}{\sqrt{1 - (\frac{1}{3})^2}} = 0.353.$$

The coordinates of event $\mathcal{A}\,(2, 1)$ in $\mathcal{K}$ have event coordinates $(1.768, 0.353)$ in $\mathcal{K}'$. We now show in Figure 9.6 the graphical solution to this problem in a spacetime diagram.

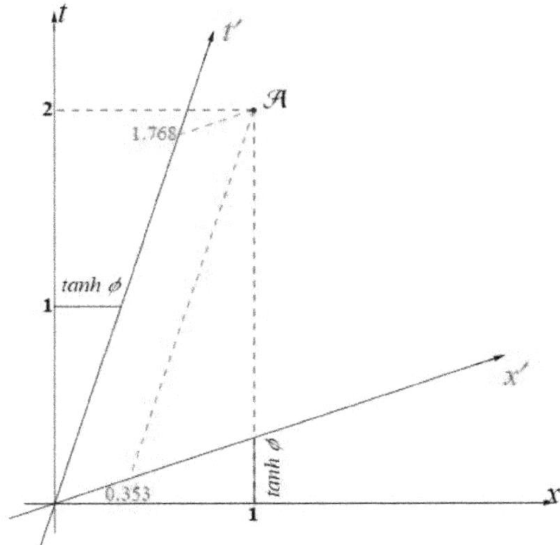

Figure 9.6: Event $\mathcal{A}$ with its coordinates in $\mathcal{K}$ and $\mathcal{K}'$.

Remark 9.3

As already mentioned above, the spacetime diagram does not represent Euclidean geometry, but Minkowskian geometry. Therefore, one must not use a ruler to measure the numerical values of Figure 9.6. In Section 9.5 we address the calibration of the axes and can thus determine the scale of the unit.

■

To transfer event intervals $(\triangle s')^2_{\mathcal{AB}}$ from the inertial frame $\mathcal{K}'$, $v = \tanh \phi$ to $\mathcal{K}$, and vice versa, it is necessary to know the conversion formulae for the respective coordinates. This is best done with a spacetime diagram and the specific hyperbola. We want to investigate a spacelike event on the x'-coordinate with distance d from the origin, and thereby determine the (t, x) coordinates with respect to the inertial frame $\mathcal{K}$ by means of the hyperbola $-(\triangle t')^2 + (\triangle x')^2 = -(\triangle t)^2 + (\triangle x)^2 = d^2$, equation (9.13). The hyperbola has in $\mathcal{K}$ the coordinates $(d \cosh \phi, d \sinh \phi)$. For this we will consider Figure 9.7. With the use of this Figure we can determine the following. The event $\mathcal{B}$ has

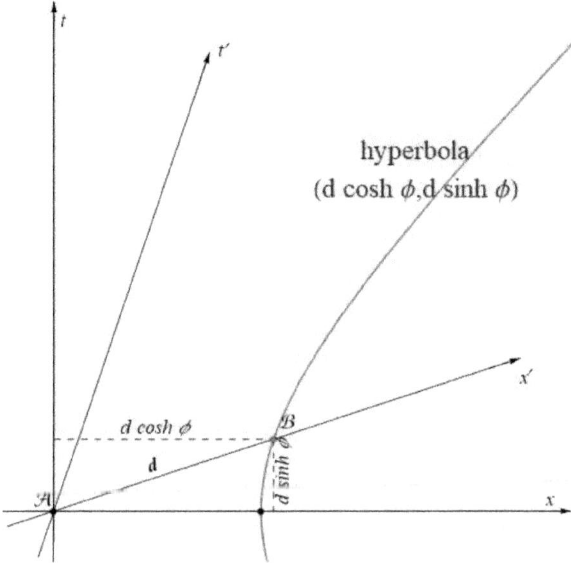

Figure 9.7: Spacetime hyperbola in the $\mathcal{K}$ spacetime diagram $(v = \tanh \phi)$.

the coordinates $(0, d)$ in $\mathcal{K}'$, and the coordinates $(d \sinh \phi, d \cosh \phi)$ in $\mathcal{K}$. And thus the following applies to the intervals $(\triangle s')^2_{\mathcal{AB}}$ and $(\triangle s)^2_{\mathcal{AB}}$:

$$(\triangle s')^2_{\mathcal{AB}} = -(\triangle t')^2 + (\triangle x')^2 = -0^2 + d^2 = d^2,$$
$$(\triangle s)^2_{\mathcal{AB}} = -(\triangle t)^2 + (\triangle x)^2$$
$$= -(d \sinh \phi)^2 + (d \cosh \phi)^2 = d^2(-\sinh^2 \phi + \cosh^2 \phi) = d^2.$$

For a timelike investigation, the timelike hyperbola must be used, with the $\mathcal{K}$ coordinates $(d \sinh \phi, d \cosh \phi)$.

9.5 Time dilation

The third postulate of SR (Summary 9.1) states that the interval is invariant with respect to any chosen inertial frame, equation (9.6). This allows us to perform a *time calibration* using the hyperbola

$$
\begin{aligned}
(\triangle s')^2 = (\triangle s)^2 &= -1, \\
-(\triangle t)^2 + (\triangle x)^2 &= -1, \\
(\triangle t)^2 - (\triangle x)^2 &= 1, \\
\cosh^2 \phi - \sinh^2 \phi &= 1.
\end{aligned}
\tag{9.27}
$$

i.e. all lightlike intervals at a distance of 1 from the origin, plotted in a space-time diagram. We look at the intersection points of the hyperbola with the t-coordinate ($x = 0$) of the inertial frame $\mathcal{K}$ and the t'-coordinate ($x' = 0$) of the inertial frame $\mathcal{K}'$. The clock in $\mathcal{K}$ at $x = 0$ displays $t = 1$, the *proper time* of $\mathcal{K}$, and the clock in $\mathcal{K}'$ at $x' = 0$ displays $t' = 1$, the *proper time* of $\mathcal{K}'$. The coordinate $t' = 1$ in $\mathcal{K}'$ has the coordinate $t = \cosh \phi = 1/\sqrt{1 - v^2}$ in $\mathcal{K}$. Figure 9.8 shows our considerations graphically. Thus the time t' in $\mathcal{K}'$ expressed

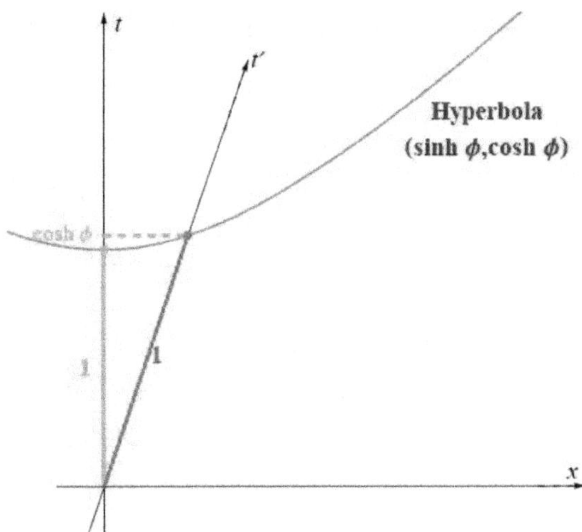

Figure 9.8: Time calibration between the inertial frames $\mathcal{K}$ and $\mathcal{K}'$ ($v = \tanh \phi$).

by the time coordinate t in $\mathcal{K}$ is stretched by the factor $\cosh \phi = \frac{1}{\sqrt{1-v^2}}$, the so-called *time dilation*.

$$(\triangle t)_{\mathcal{K}} = \frac{1}{\sqrt{1 - v^2}} (\triangle t)_{\mathcal{K}'},$$

$$dt = \frac{1}{\sqrt{1 - v^2}} d\tau, \qquad (9.28)$$

$$dt = \gamma \, d\tau.$$

Therefore, clocks in a moving inertial frame $\mathcal{K}'$ ($v = \tanh \phi$) run slower compared to clocks at rest.

9.6 Length contraction

In most textbooks on SR, one finds a very simple derivation of the physical phenomenon of *length contraction*, namely that a rod, moving in the direction of its length with velocity v, appears to the resting observer to be shortened in its longitudinal extension. Let us briefly reproduce such a derivation. One uses only one of the two equations of the LT, equation (9.21), namely: $x' = \gamma(x - vt)$. The measuring process takes place in the inertial frame $\mathcal{K}$ and the both ends of the rod are measured at time $t_1 = t_2$. Therefore, the following applies to the rod length $l_{\mathcal{K}'}$:

$$\begin{aligned}
l_{\mathcal{K}'} &= x_2' - x_1' \\
&= \gamma(x_2 - vt_2) - \gamma(x_1 - vt_1) \\
&= \gamma(x_2 - x_1) \\
&= \gamma l_{\mathcal{K}}.
\end{aligned}$$

The result is correct, but the second equation concerning the time transformation of the LT was not taken into account. If the rod is at rest in $\mathcal{K}'$, then $t_1' = t_2' = 0$. But during the measurement process in $\mathcal{K}$ with event (t_2, x_2), t_2' is not equal to zero. An exact solution of this task is possible by means of a spacetime diagram.

In order to take the time coordinate t' of $\mathcal{K}'$ into account, it is best to use a $\mathcal{K}$ spacetime diagram (see Figure 9.9) and take the results from the $\mathcal{K}$ spacetime diagram of Figure 9.7. The *proper length* is defined as the length that an object has when it is at rest. The rod $\mathcal{AB}$, which rest in $\mathcal{K}'$, has the proper

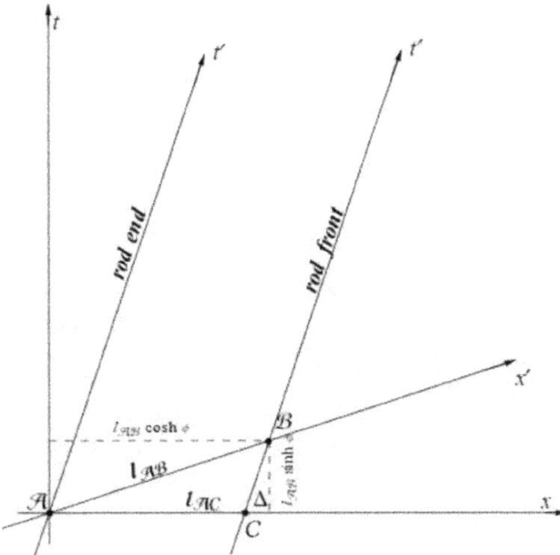

Figure 9.9: Length comparison of a rod l_{AB} resting in $\mathcal{K}'$ to its measured length l_{AC} in $\mathcal{K}$.

length l_{AB}. To include the time component of event B in the calculation, the time coordinate of the rod front t'_B must be included in the spacetime diagram (see Figure 9.9). When measuring this length l_{AB} in another inertial frame, e.g. $\mathcal{K}$, it must be noted that the measurement of the ends of the rod occurs at an identical time $t_A = t_C$, e.g. $t = 0$. For this reason the intersection points of the time coordinates t'_A, t'_B of A, B with the x-coordinate ($t_A = t_C = 0$) of $\mathcal{K}$ must be determined.

With these preliminary considerations, the following can be determined with the use of the spacetime diagram, Figure 9.9: The measured length l_{AC} in $\mathcal{K}$ is

$$l_{AC} = l_{AB} \cosh \phi - \triangle. \tag{9.29}$$

The slope of the t'-coordinate with respect to the t-coordinate of $\mathcal{K}$ is $v = \tanh \phi = \triangle x / \triangle t$. And for $t_{B'}$, therefore: $\tanh \phi = \triangle x / \triangle t = \triangle / l_{AB} \sinh \phi$. And thus

$$
\begin{aligned}
l_{AC} &= l_{AB} \cosh \phi - \triangle \\
&= l_{AB} \cosh \phi - l_{AB} \sinh \phi \tanh \phi \\
&= l_{AB} \frac{\cosh^2 \phi - \sinh^2 \phi}{\cosh \phi} \\
&= \frac{l_{AB}}{\cosh \phi} = l_{AB} \sqrt{1 - v^2}.
\end{aligned}
\tag{9.30}
$$

This is the so-called *Lorentz contraction* or *length contraction*.

9.7 Vectors

Remark 9.4
1. Vectors, regardless of their dimensionality, are marked with an arrow above their designation. If there can be confusion between a 4-vector of SR and a 3-vector of ordinary spatiality, this is specifically indicated.
2. It is always important to note whether one is dealing with a vector with *contravariant* components (superscript index) or a covector with *covariant* components (subscript index). Contravariant components are usually represented as column vectors:

$$\vec{x} = \begin{pmatrix} x^1 \\ x^2 \\ x^3 \end{pmatrix}.$$

For reasons of space, we will represent them in a row with a superscript T (T for transposed):

$$\vec{x} = \begin{pmatrix} x^1 & x^2 & x^3 \end{pmatrix}^T.$$

Covariant components are summarised in one row as usual:

$$\vec{x} = \begin{pmatrix} x_1 & x_2 & x_3 \end{pmatrix}.$$

■

In this section we want to introduce the so-called 4-*vector* of the SR. Particular features (change from contravariant representation to covariant, inner product) occur due to the special metric η of the Minkowski space MS. In Section 9.2, (9.7), we have already learned about the canonical metric tensor η_{ij} of MS:

$$\eta = \begin{pmatrix} -1 & 0 & 0 & 0 \\ 0 & 1 & 0 & 0 \\ 0 & 0 & 1 & 0 \\ 0 & 0 & 0 & 1 \end{pmatrix}, \; \eta_{ij} = \begin{cases} -1 & \text{for } i = j = 0 \\ 1 & \text{for } i = j = 1, 2, 3 \\ 0 & \text{for } i \neq j \end{cases}$$

The scalar product of the basis vectors of MS $\{\vec{e}_0, \vec{e}_1, \vec{e}_2, \vec{e}_3\}$ and the components of η are related as follows:

$$\eta_{ij} = \vec{e}_i \cdot \vec{e}_j, \text{ for } i, j = 0, 1, 2, 3,$$
$$\Longrightarrow$$
$$\vec{e}_0 \cdot \vec{e}_0 = -1,$$
$$\vec{e}_i \cdot \vec{e}_i = 1, \text{ for } i = 1, 2, 3.$$

(9.31)

In MS, the general rules of a vector space apply, see Chapter 1. A general vector $\vec{x}$ in $\mathcal{K}$, a so-called 4-*vector*, and its differential in component representation with respect to the origin is defined as

$$\begin{aligned}
\vec{x} &= (x^0, \vec{r})^T \\
&= x^0 \vec{e}_0 + x^1 \vec{e}_1 + x^2 \vec{e}_2 + x^3 \vec{e}_3 = (x^0, x^1, x^2, x^3)^T, \\
&= t\vec{e}_0 + x\vec{e}_x + y\vec{e}_y + z\vec{e}_z = (t, x, y, z)^T, \\
d\vec{x} &= (dx^0, dx^1, dx^2, dx^3)^T, \\
&= (dt, dx, dy, dz)^T,
\end{aligned}$$

(9.32)

with $\vec{r}$ as the usual 3-dimensional space vector.

Example 9.2

x^+-boost Lorentz transformation of a given 4-vector $\vec{a} = \left(a^0, a^1, a^2, a^3\right)^T = (3, 0, 2, 1)$.

With equation 9.21 we obtain

$$\vec{a}' = L\,\vec{a},$$

$$\begin{pmatrix} a^{0'} \\ a^{1'} \\ a^{2'} \\ a^{3'} \end{pmatrix} = \begin{pmatrix} \gamma & -\gamma v & 0 & 0 \\ -\gamma v & \gamma & 0 & 0 \\ 0 & 0 & 1 & 0 \\ 0 & 0 & 0 & 1 \end{pmatrix} \begin{pmatrix} a^0 \\ a^1 \\ a^2 \\ a^3 \end{pmatrix}$$

$$= \begin{pmatrix} \gamma & -\gamma v & 0 & 0 \\ -\gamma v & \gamma & 0 & 0 \\ 0 & 0 & 1 & 0 \\ 0 & 0 & 0 & 1 \end{pmatrix} \begin{pmatrix} 3 \\ 0 \\ 2 \\ 1 \end{pmatrix} = \begin{pmatrix} 3\gamma \\ -3\gamma v \\ 2 \\ 1 \end{pmatrix}.$$

∎

The *scalar product* or *inner product* of two 4-vectors and the *magnitude* of a 4-vector is defined using the canonical metric tensor of Minkowski space, equation (9.7). The scalar product of 4-vectors $\vec{a}$ and $\vec{b}$ equals:

$$
\begin{aligned}
\vec{a} &= \left(a^0, a^1, a^2, a^3\right)^T, \\
\vec{b} &= \left(b^0, b^1, b^2, b^3\right)^T, \\
\vec{a} \cdot \vec{b} &= \langle \vec{a}, \vec{b} \rangle = \eta_{ij} a^i b^j \\
&= -a^0 b^0 + a^1 b^1 + a^2 b^2 + a^3 b^3.
\end{aligned}
\tag{9.33}
$$

If we apply the lowering of an index by the metric tensor η_{ij}, (9.7), we obtain a contraction over the indices of covariant (one-form) and contravariant (vector) components.

$$
\vec{a} \cdot \vec{b} = \eta_{ij} a^i b^j = a_j b^j.
\tag{9.34}
$$

We have to keep in mind that the sign changes when the index $i = 0$ is lowered or raised.

$$
a_j = \eta_{0j} a^0, \quad \eta_{0j} = \begin{cases} -1 & j = 0 \\ 0 & j \neq 0 \end{cases}
$$

$$
\Longrightarrow
$$

$$
a_0 = -a^0.
\tag{9.35}
$$

And again applied to the scalar product:

$$
\begin{aligned}
\vec{a} \cdot \vec{b} &= a_j b^j \\
&= a_0 b^0 + a_1 b^1 + a_2 b^2 + a_3 b^3 \\
&= -a^0 b^0 + a^1 b^1 + a^2 b^2 + a^3 b^3.
\end{aligned}
\tag{9.36}
$$

The magnitude or *norm* of 4-vector $\vec{A}$ is:

$$
||\vec{a}|| = \vec{a} \cdot \vec{a} = \eta_{ij} a^i a^j = -(a^0)^2 + (a^1)^2 + (a^2)^2 + (a^3)^2.
\tag{9.37}
$$

As with event intervals, the magnitude of a 4-vector can be negative, zero or positive. One also adopts the designation of the intervals for the 4-vectors: A 4-vector with a negative magnitude is a *timelike vector*, if the magnitude is zero it is a *null vector* and if the magnitude is positive it is a *spacelike vector*.

Box 9.4

The magnitude of a 4-vector is invariant under the Lorentz transformation, i.e. it is a frame invariant number.

In Box 9.3 we have already shown that any event interval is invariant under LT. Of course, this also applies to the magnitude of a 4-vector. In the expressions of Box 9.3 one replaces the differentials of the interval by the components of the 4-vector, and the proof follows stringently.

We want to verify this with the concrete 4-vector $\vec{a} = (3, 0, 2, 1)$ from Example 9.2.

$$||\vec{a}|| = -(a^0)^2 + (a^1)^2 + (a^2)^2 + (a^3)^2$$
$$= -3^2 + 0^2 + 2^2 + 1^2 = -4.$$

We use the more straightforward hyperbolic components of the LT matrix, $\gamma = \cosh\phi$, $v\gamma = \sinh\phi$, and obtain for $\vec{a}'$ from Example 9.2:

$$\vec{a}' = (3\cosh\phi, -3\sinh\phi, 2, 1).$$

And the magnitude of $\vec{a}'$ is now given by:

$$\left\|\vec{a}'\right\| = -(3\cosh\phi)^2 + (-3\sinh\phi)^2 + 2^2 + 1^2$$
$$= -9\underbrace{(\cosh^2\phi - \sinh^2\phi)}_{=1} + 4 + 1 = -4.$$

At this point we would like to summarise the transformation rules for the components of a 4-vector and those of the basis vectors. In a x^+-boost Lorentz transformation ($\mathcal{K} \longrightarrow \mathcal{K}'$), the components x^j of an event vector $\vec{x}$ of $\mathcal{K}$ transform according to the Lorentz transformation matrix $L^{i'}_j$, Section 9.3. (9.21), as follows:

$$x^j \text{ of } \mathcal{K} \xrightarrow{L^{i'}_j} x^{i'} \text{ of } \mathcal{K}'$$
$$x^{i'} = L^{i'}_j\, x^j, \tag{9.38}$$

with

$$L^{i'}_{j} = \begin{pmatrix} \gamma & -\gamma v & 0 & 0 \\ -\gamma v & \gamma & 0 & 0 \\ 0 & 0 & 1 & 0 \\ 0 & 0 & 0 & 1 \end{pmatrix} = \begin{pmatrix} \cosh\phi & -\sinh\phi & 0 & 0 \\ -\sinh\phi & \cosh\phi & 0 & 0 \\ 0 & 0 & 1 & 0 \\ 0 & 0 & 0 & 1 \end{pmatrix}, v = \tanh\phi.$$

As we know from Chapter 1, the components of a vector transform contravariant, and the basis vectors covariant, which will be briefly shown again below. An event 4-vector $\vec{x}$ represents a physical event and thus invariant to a frame transition:

$$\vec{x} = x^{i'}\vec{e}_{i'} = x^{j}\vec{e}_{j},$$

$$L^{i'}_{j} x^{j}\vec{e}_{i'} = x^{j}\vec{e}_{j},$$

$$\Longrightarrow$$

$$L^{i'}_{j}\vec{e}_{i'} = \vec{e}_{j}.$$

(9.39)

It can be seen from the comparison of equations (9.38) and (9.39) that the transformation behaviour of basis vectors and components of vectors in MS are inverse to each other under the Lorentz transformation $\mathcal{K} \xrightarrow{L^{i'}_{j}} \mathcal{K}'$. We now want to show this more clearly.

$$L^{i'}_{j}\vec{e}_{i'} = \vec{e}_{j},$$

$$(L^{i'}_{j})^{-1}L^{i'}_{j}\vec{e}_{i'} = (L^{i'}_{j})^{-1}\vec{e}_{j},$$

$$\vec{e}_{i'} = L^{j}_{i'}\vec{e}_{j}$$

(9.40)

with

$$L^{j}_{i'} = \begin{pmatrix} \gamma & \gamma v & 0 & 0 \\ \gamma v & \gamma & 0 & 0 \\ 0 & 0 & 1 & 0 \\ 0 & 0 & 0 & 1 \end{pmatrix} = \begin{pmatrix} \cosh\phi & \sinh\phi & 0 & 0 \\ \sinh\phi & \cosh\phi & 0 & 0 \\ 0 & 0 & 1 & 0 \\ 0 & 0 & 0 & 1 \end{pmatrix}, v = \tanh\phi.$$

9.8 Momentum and Energy

In 3-dimensional geometry, velocity is a tangent vector to the path of a particle. We are now looking for the velocity vector of a particle in 4-dimensional Minkowski space. As explained in Section 9.2, the curves of massed objects are timelike world lines. To study the equations of motion of such objects, we use the method of parametrisation of curves $x^i(\lambda)$, Section 7.2, specified by a mapping $\mathbb{R} \longrightarrow M$, where M is the manifold representing the Minkowski space. As a parameter we use the proper time $d\tau$, (9.9), which is invariant under the Lorentz transformation. The *tangential vector*

$$u^i = \frac{dx^i(\tau)}{d\tau} \tag{9.41}$$

is a 4-vector called 4-*velocity*. Because of $d\tau^2 = -\eta_{ij}dx^i dx^j$, (9.9), the four-velocity u^i is automatically normalized:

$$
\begin{aligned}
-d\tau^2 &= \eta_{ij}dx^i dx^j, \\
-1 &= \eta_{ij}\frac{dx^i}{d\tau}\frac{dx^j}{d\tau} \\
&= \eta_{ij}u^i u^j = ||\vec{u}|| .
\end{aligned}
\tag{9.42}
$$

Therefore, for a particle at rest ($x^{i'} = 0$) in $\mathcal{K}'$, its 4-velocity is:

$$
\begin{aligned}
\vec{u}' &= (1,0,0,0)^T = 1 \cdot \vec{e}_{0'}, \\
\vec{u}' \cdot \vec{u}' &= \vec{e}_{0'} \cdot \vec{e}_{0'} = -1.
\end{aligned}
\tag{9.43}
$$

A particle at rest in $\mathcal{K}'$ means that it is firmly fixed at the origin $\mathcal{O}'$ and takes its coordinate system $\mathcal{K}'$ with it. We therefore call it a *co-moving reference frame* CRF. In this coordinate system, the 4-velocity $\vec{u}'$ of this particle is always $\vec{u}' = \vec{e}_{0'}$, parallel to the time axis with magnitude -1. Due to the restriction of the parametrisation by the proper time τ, the 4-vector $\vec{u}'$ describes a geodesic. The tangent $\vec{e}_{0'}$ of the timelike curve is parallel transported. The geodesic generated in this way is the world line of the particle. Let us graphically illustrate what we have just described, Figure 9.10.

And the 4-velocity $\vec{u}'$ deduced from the 4-event vector $\vec{x}'$ of a particle in his rest frame $\mathcal{K}'$ (CRF):

$$\vec{x}' = (x^{0'}, x^{1'}, x^{2'}, x^{3'})^T$$
$$= (\tau, 0, 0, 0)^T,$$
$$d\vec{x}' = (d\tau, 0, 0, 0)^T, \qquad (9.44)$$
$$\Longrightarrow$$
$$\frac{d\vec{x}'}{d\tau} = \vec{u}' = (1, 0, 0, 0)^T.$$

The frame $\mathcal{K}'$, the CRF of a particle, moves with velocity v relative to the frame

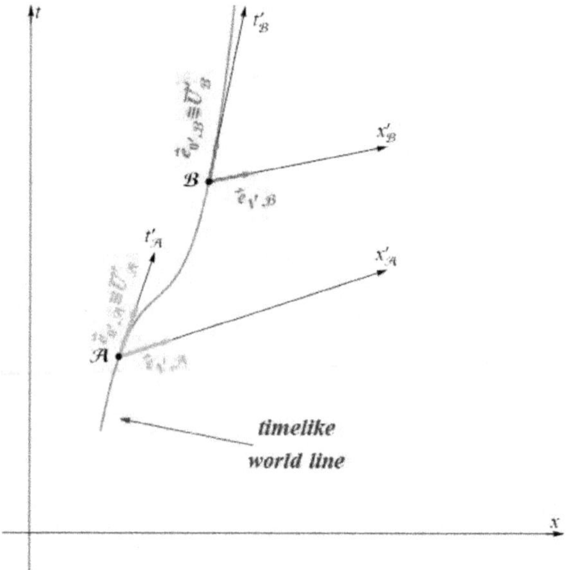

Figure 9.10: CRF basis vectors with the corresponding 4-velocity at two events $\mathcal{A}$ and $\mathcal{B}$ of the world line.

$\mathcal{K}$ in the positive x-direction. Thus, the velocity of the particle is identical to that of $\mathcal{K}'$. This leads us to the components of the 4-velocity $\vec{u}$ by applying the LT, equation (9.20):

$$\vec{u} = L\vec{u}' = \begin{pmatrix} \gamma & \gamma v & 0 & 0 \\ \gamma v & \gamma & 0 & 0 \\ 0 & 0 & 1 & 0 \\ 0 & 0 & 0 & 1 \end{pmatrix} \begin{pmatrix} 1 \\ 0 \\ 0 \\ 0 \end{pmatrix} = \begin{pmatrix} \gamma \\ \gamma v \\ 0 \\ 0 \end{pmatrix}. \qquad (9.45)$$

Thus we have obtained the 4-velocity vector $\vec{u}$ of a particle in *spacetime* moving with velocity v in the positive x-direction

$$\vec{u} = (\gamma, \gamma v, 0, 0)^T. \tag{9.46}$$

We can also receive this result directly via the differential $u^i = \frac{dx^i}{d\tau}$, (9.41), using the 4-event differential, (9.32), and the time dilation, (9.28):

$$u^i = \frac{dx^i}{d\tau},$$

for $i = 0$:

$$u^0 = \frac{dx^0}{d\tau} = \frac{dt}{d\tau} = \gamma,$$

for $i = 1, 2, 3$:

$$\left(\frac{dx}{d\tau}, \frac{dy}{d\tau}, \frac{dz}{d\tau}\right)^T = \gamma \left(\frac{dx}{dt}, \frac{dy}{dt}, \frac{dz}{dt}\right)^T = \gamma\vec{v},$$

and therefore

$$\vec{u} = \begin{pmatrix} \gamma \\ \gamma\vec{v} \end{pmatrix}, \tag{9.47}$$

where $\vec{v}$ is the ordinary 3-spatial velocity. It is quite easy to calculate that $||\vec{u}|| = -1$.

Remark 9.5
The 4-velocity is a velocity in *spacetime* and *not* the ordinary 3-component velocity of space. The 4-velocity is a tangent vector to the world line. The world line itself, parametrised by the proper time $\frac{d}{d\tau}$, is a geodesic. Section 7.6, (7.12) with $u^i = \frac{dx^i}{d\tau}$.

■

The 4-*momentum vector* $\vec{p}$ of the Minkowski space is defined according to the classical definition of momentum by

$$\vec{p} = m\,\vec{u}, \tag{9.48}$$

where m is the mass of the particle measured when it is at rest, for example in its CRF, and $\vec{u}$ is the 4-velocity. With this information we already get a first important and fundamental relationship. The 4-momentum $\vec{p}\,'$ of a particle in the CRF is therefore:

$$\vec{p}\,' = m\,\vec{u}\,'$$
$$= m\,(1,0,0,0)^T, \tag{9.49}$$
$$p^{0'} = m.$$

The energy E of a particle is identified with the $i = 0$ component of the 4-momentum vector. For the energy of a particle at rest is its rest energy $E = m\,(c = 1)$, or $E = m\,c^2$. In this way we have derived Albert Einstein's famous formula.

The 4-momentum vector $\vec{p}$ of a particle moving with velocity v is according to the Lorentz transformation (9.20),

$$\vec{p} = L\,\vec{p}\,' = \begin{pmatrix} \gamma & \gamma v & 0 & 0 \\ \gamma v & \gamma & 0 & 0 \\ 0 & 0 & 1 & 0 \\ 0 & 0 & 0 & 1 \end{pmatrix} \begin{pmatrix} m \\ 0 \\ 0 \\ 0 \end{pmatrix} = \begin{pmatrix} \gamma m \\ \gamma v m \\ 0 \\ 0 \end{pmatrix}. \tag{9.50}$$

Let us consider the classical limiting case $v \ll 1$:

$$\gamma = 1 + \frac{1}{2}v^2 + \mathcal{O}(v^4),$$
$$\gamma v = v + \mathcal{O}(v^3).$$

For the energy $E(i = 0)$ this results in

$$p^0 = \gamma m \longrightarrow \underbrace{m}_{\text{rest energy}} + \underbrace{\frac{1}{2}m\,v^2}_{\text{kinetic energy}},$$

and for the momentum components $p^i\,(i = 1,2,3)$ we obtain in the limiting case

$$p^1 = \gamma\,v\,m \longrightarrow m\,v.$$

This is again a fine example of how classical Newtonian physics emerges as a limiting case of SR.

What do we get when we examine the norm or scalar product of the 4-momentum vector? A first important result is that the norm of the 4-momentum is always equal to $-m^2$, regardless of whether the particle is in its rest frame CRF or is moving in the inertial frame $\mathcal{K}$ with velocity v (Invariance of the scalar product or norm under the Lorentz transformation, Box 9.3

and Box 9.4).

For CRF we get, (9.49):

$$\vec{p}\,' \cdot \vec{p}\,' = -(p^{0'})^2 = -m^2.$$

And in a $\mathcal{K}$ frame (x^+-boost, equation (6.12)) we obtain

$$\vec{p} \cdot \vec{p} = -(p^0)^2 + (p^1)^2 = -(\gamma m)^2 + (\gamma v m)^2 = -m^2.$$

This is the law of *conservation of energy*. And as a second result for $\vec{p} \cdot \vec{p}$ with $p^0 = E$ we get the following relationship

$$\begin{aligned}
\vec{p} \cdot \vec{p} &= -(p^0)^2 + (p^1)^2 + (p^2)^2 + (p^3)^2 \\
&= -E^2 + \underbrace{(\gamma m v^1)^2 + (\gamma m v^2)^2 + (\gamma m v^3)^2}_{=\boldsymbol{p}^2,\ \boldsymbol{p}=\gamma m v^i,\ \text{the spatial 3-momentum}} \\
&= -m^2,
\end{aligned}$$

and therefore

$$\boxed{E^2 = m^2 + \boldsymbol{p}^2} \tag{9.51}$$

with $\boldsymbol{p} = \gamma m v^i = \gamma m \vec{v}$, the spatial 3-momentum.

The 4-*force vector* $\vec{f}$, the Minkowski force, following its Newtonian classical model $\boldsymbol{F} = m\,\boldsymbol{a}$, is defined as

$$\begin{aligned}
\vec{f} &= m\,\vec{a} = m\,\frac{d\vec{u}}{d\tau}, \\
f^i &= m\,a^i = m\,\frac{du^i}{d\tau} = m\,\frac{d^2 x^i}{d\tau^2},
\end{aligned} \tag{9.52}$$

where $\vec{a}$ is the 4-*acceleration vector*. Thus, the principle that a force-free particle remains at rest or in uniform motion also applies in ST. But the *gravitational* force, which is usually processed in classical mechanics using Newton's formula, cannot do so in relativistic mechanics using the formula just described, (9.52). In relativistic mechanics, gravity is not a force but a *curvature* of 4-dimensional spacetime.

We now want to point out an important relationship between the 4-velocity and the 4-acceleration or 4-force. We consider a particle in its rest frame, the CRF. Then, according to (9.43)

$$\vec{u}' = (1,0,0,0)^T = 1 \cdot \vec{e}_{0'},$$
$$\vec{u}' \cdot \vec{u}' = \vec{e}_{0'} \cdot \vec{e}_{0'} = -1,$$

we obtain with differentiation with respect to proper time τ

$$\vec{u}' \cdot \vec{u}' = -1,$$
$$\frac{d}{d\tau}(\vec{u}' \cdot \vec{u}') = 0, \tag{9.53}$$
$$\vec{u}' \cdot \frac{d\vec{u}'}{d\tau} = 0.$$

Since $\vec{u}'$ has only the $i = 0$ component (all others are zero), $\frac{d\vec{u}'}{d\tau}$ is constrained to have only *spatial components*

$$\frac{d\vec{u}'}{d\tau} = (0, a^1, a^2, a^3)^T = \vec{a}', \tag{9.54}$$

and therefore '

$$\vec{u}' \cdot \vec{a}' = 0. \tag{9.55}$$

Since a scalar product in MS is invariant under the Lorentz transformation, equation (9.55) is also valid for the particle in the $\mathcal{K}$ frame with $\vec{u} = \begin{pmatrix} \gamma \\ \gamma\vec{v} \end{pmatrix}$:

$$\begin{aligned}
\vec{u}' \cdot \vec{a}' &= \eta_{i'j'} u^{i'} a^{j'} \\
&= \eta_{i'j'} \left(L^{i'}_i u^i \right) \left(L^{j'}_j a^j \right) \\
&= \left(L^{i'}_i L^{j'}_j \eta_{i'j'} \right) u^i a^j \\
&= \eta_{ij} u^i a^j = \vec{u} \cdot \vec{a}.
\end{aligned} \tag{9.56}$$

The 4-velocity vector and the 4-acceleration vector are always orthogonal to each other in Minkowski space.

Example 9.3

Electrodynamics: 3-dimensional Lorentz force versus 4-dimensional Minkowski force

The well known 3-dimensional Lorentz force of a moving charge q in an electromagnetic field, $\vec{F} = q\,(\vec{E} + \vec{v} \times \vec{B})$, is replaced in the relativistic representation of electrodynamics by the so-called 4-dimensional Minkowski force, which states as follows

$$f^i = q\,u_j F^{ij}, \qquad (9.57)$$

where F^{ij} is the electromagnetic field tensor

$$F^{ij} = \begin{pmatrix} 0 & E^1 & E^2 & E^3 \\ -E^1 & 0 & B^3 & -B^2 \\ -E^2 & -B^3 & 0 & B^1 \\ -E^3 & B^2 & -B^1 & 0 \end{pmatrix}, \; i,j = 0,1,2,3, \; F^{(i+1=\text{row})(j+1=\text{column})},$$

and u_j is a covariant component of the 4-velocity $\vec{u}$: $u_j = \eta_{jk} u^k$, with

$$\vec{u} = \begin{pmatrix} \gamma \\ \gamma\vec{v} \end{pmatrix} = \begin{pmatrix} \gamma \\ \gamma v^1 \\ \gamma v^2 \\ \gamma v^3 \end{pmatrix}.$$

For $i = 1$, the *x-spatial* component results in

$$\begin{aligned}
f^1 &= q\,\eta_{jk}\,u^k\,F^{1j} \\
&= q\,\left(\eta_{00} u^0 F^{10} + \eta_{11} u^1 F^{11} + \eta_{22} u^2 F^{12} + \eta_{33} u^3 F^{13} \right) \\
&= q\,\left(-u^0(-E^1) + 0 + u^2 B^3 + u^3(-B^2) \right) \\
&= q\,\left(\gamma E^1 + \gamma v^2 B^3 - \gamma v^3 B^2 \right) \\
&= \gamma q\,\left(\vec{E} + \vec{v} \times \vec{B} \right)^1 .
\end{aligned}$$

The *spatial* component of the 4-Minkowski force is

$$\vec{f}_{\text{spatial}} = \gamma q\,\left(\vec{E} + \vec{v} \times \vec{B} \right) \qquad (9.58)$$

and corresponds to the known non-relativistic Lorentz force with the additional relativistic factor γ.

For $i = 0$ this results in

$$
\begin{aligned}
f^0 &= q\,\eta_{jk}\,u^k\,F^{0j} \\
&= q\left(\eta_{00}u^0 F^{00} + \eta_{11}u^1 F^{01} + \eta_{22}u^2 F^{02} + \eta_{33}u^3 F^{03}\right) \\
&= q\left(0 + u^1 E^1 + u^2 E^2 + u^3 E^3\right) \\
&= q\left(\gamma v^1 E^1 + \gamma v^2 E^2 + \gamma v^3 E^3\right) \\
&= \gamma q\,\vec{v}\cdot\vec{E}.
\end{aligned}
\tag{9.59}
$$

The "time" component of $\vec{f}$ corresponds to the *power* absorbed in an electric field $\vec{E}$ by the charge q. In summary, we can represent the 4-dimensional Minkowski vector as follows

$$
\vec{f} = \gamma\,q\,\begin{pmatrix} \vec{v}\cdot\vec{E} \\ \vec{E} + \vec{v}\times\vec{B} \end{pmatrix}.
\tag{9.60}
$$

∎

Appendix A

Some mathematical terms

affine parameter

An affine parameter λ in a formula leads to the same result with a linear transformation $\lambda \longrightarrow a\lambda + b$. For example see geodesic equation (7.13).

affine transformation (Euclidean geometry)

An affine transformation of the Euclidean geometry is a linear transformation that keeps distance ratios and angles constant.

affine transformation (Physics)

Physical laws apply regardless of the coordinate system. Transformations in this sense are *affine transformations* and leave for example distances invariant (e.g. Lorentz transformation).

bilinear transformation

A *bilinear transformation* $f(\ ,\ )$ is a mapping with the following property. Let $\{x, y, \ldots\}$ be variables of a well-defined domain and $a, b, \ldots$ scalars, then holds:

$$f(a\,x + b\,y,\ z) = af(x, z) + bf(y, z).$$

Of course, what has been shown also applies to the second argument, or any mixed application.

Some examples: 2-forms, metric tensor $\mathbf{g}$.

covariant

The representation of a physical correlation of physical quantities (example: $\vec{f} = m\,\vec{a}$) should be independent of the chosen reference system. Mathematical operations, transformations and physical quantities that fulfil this condition are called *covariant*. These are described by the use of tensors: scalar $(0,0)$-tensor, vector $(1,0)$-tensor, metric $(0,2)$-tensor, etc.

diffeomorph

Diffeomorphic mappings are differentiable (generally infinitely differentiable C^∞), bijective mappings f for which both f and f^{-1} are *smooth*.

group

A set G of elements $\{g_i\}$ together with a binary operation $\circ$ is a group if the following three axioms are fulfilled:

1. associative: $g_1 \circ (g_2 \circ g_3) = (g_1 \circ g_2) \circ g_3$.

2. identity: there exists an $e \in G$ such that $e \circ g_i = g_i \circ e = g_i$, $\forall i$.

3. inverse: for each g_i there is an g_i^{-1} such that: $g_i \circ g_i^{-1} = g_i^{-1} \circ g_i = e$.

A group is called Abelian (commutative) if

4. $g_i \circ g_j = g_j \circ g_i$.

homeomorph

Homeomorphic mappings are bijective mappings $f : A \longrightarrow B$, for which both f and f^{-1} are *continuous*.

inertial frame

An *inertial frame* $\mathcal{K}$ is characterised by the fact that a force-free particle moves in a straight line (geodesics) and at constant speed. Every frame $\mathcal{K}'$ moving with constant velocity relative to $\mathcal{K}$ is also an inertial frame.

linear transformation

A *linear transformation* $f(\ )$ is a mapping with the following property. Let $\{x, y, \ldots\}$ be variables of a well-defined domain and $a, b, \ldots$ scalars, then holds:

$$f(a\,x + b\,y) = af(x) + bf(y).$$

Some examples: matrices, one-forms, differential operator $\frac{d}{dx}$.

order $\mathcal{O}$

The symbol $\mathcal{O}(\delta^n)$ means the terms of order δ^n and higher powers of an expression.

Appendix B

Determinants and Square Matrices

B.1 Determinants

A *determinant* **det** is a functional that maps any square matrix to a number. The determinant of a matrix A is denoted $\det A$.

$$\det : \mathbb{R}^{n \times n} \longrightarrow \mathbb{R}. \tag{B.1}$$

We assume the following matrix nomenclature: the upper index designates the row, the lower index the column. A be an $n \times n$ matrix:

$$A = \left(a^i_j\right) = \begin{pmatrix} a^1_1 & a^1_2 & \cdots & a^1_n \\ a^2_1 & a^2_2 & \cdots & \vdots \\ \vdots & \vdots & \ddots & \vdots \\ a^n_1 & \cdots & \cdots & a^n_n \end{pmatrix}, \tag{B.2}$$

and

$$\det\left(A\right) = \det \begin{pmatrix} a^1_1 & a^1_2 & \cdots & a^1_n \\ a^2_1 & a^2_2 & \cdots & \vdots \\ \vdots & \vdots & \ddots & \vdots \\ a^n_1 & \cdots & \cdots & a^n_n \end{pmatrix} = \begin{vmatrix} a^1_1 & a^1_2 & \cdots & a^1_n \\ a^2_1 & a^2_2 & \cdots & \vdots \\ \vdots & \vdots & \ddots & \vdots \\ a^n_1 & \cdots & \cdots & a^n_n \end{vmatrix}. \tag{B.3}$$

The **det**(A) functional is subject to the following rules:

1. Interchanging two rows or columns does not change the value of the determinant, but it does change the sign.

2. Transposing does not change the value or the sign.

3. If each entry in a row or column is multiplied by a factor r, the determinant is multiplied by r.

4. If two rows or columns are identical, the determinant has the value zero. Likewise, if two rows (columns) are proportional to each other, the value is zero.

5. A multiple of a row (column) can be added to any other row (column) without changing the value of the determinant.

6. If a row (column) consists only of zero entries, the determinant is zero.

7. $\det(A)$ is a linear functional in the rows (columns). Example:

$$\begin{vmatrix} \vdots & \vdots & \vdots \\ 1+2 & 4+5 & 3+1 \\ \vdots & \vdots & \vdots \end{vmatrix} = \begin{vmatrix} \vdots & \vdots & \vdots \\ 1 & 4 & 3 \\ \vdots & \vdots & \vdots \end{vmatrix} + \begin{vmatrix} \vdots & \vdots & \vdots \\ 2 & 5 & 1 \\ \vdots & \vdots & \vdots \end{vmatrix}.$$

All these constraints are summarised in the so-called *complete expansion* of the determinant of an $n \times n$ matrix with elements $a^i_j (i, j = 1, \ldots, n)$:

$$\det(A) = \sum_{i_1 i_2 \ldots i_n} \epsilon_{i_1 i_2 \ldots i_n}\, a^1_{i_1} a^2_{i_2} \ldots a^n_{i_n}, \quad \text{(selected rows)},$$

$$= \epsilon_{i_1 i_2 \ldots i_n}\, a^{i_1}_1 a^{i_2}_2 \ldots a^{i_n}_n, \quad \text{(selected columns)},$$

(B.4)

with the Levi-Civita symbol

$$\epsilon_{i_1 i_2 \ldots i_n} = \begin{cases} +1, & i_1 i_2 \ldots i_n \text{ is an even permutation of } 12 \ldots n, \\ -1, & i_1 i_2 \ldots i_n \text{ is an odd permutation of } 12 \ldots n, \\ 0 & i_1 i_2 \ldots i_n \text{ is not a permutation of } 12 \ldots n. \end{cases}$$

(B.5)

In the first line of (B.4) we have included the sum sign to make it clear that two subscript index sets are being summed. For the term 'permutation' see Box 4.1.

The *Laplace expansion* is a recursive method in which a determinant of order n is expressed by n determinants of order $(n-1)$. A reduced matrix, $(n-1) \times (n-1)$, is formed from a $(n \times n)$-matrix A by removing the i-th row and the j-th column. The determinant of this "sub-matrix" is called *minor* and is denoted M^i_j. For any fixed i:

$$\det(A) = \sum_j a^i_j (-1)^{i+j} M^i_j = \sum_j a^i_j C^i_j. \tag{B.6}$$

C^i_j is called *cofactor* of the matrix element a^i_j.

Using the Laplace expansion equation (B.6) to determine the determinant of an $(n \times n)$-matrix, we can derive a formula for obtaining the inverse of a matrix. We proceed as follows:

$$\det(A) = \sum_j a^i_j C^i_j, \text{ for any fixed } i,$$

$$1 = \sum_j a^i_j \frac{C^i_j}{\det(A)}, \text{ for any fixed } i.$$

With $(A^{-1})^j_i = \frac{C^i_j}{\det(A)}$ and the index summation rule for matrices follows

$$1 = a^i_j (A^{-1})^j_i, \text{ for any fixed } i.$$

The formula for the inverse of a matrix is therefore:

$$(A^{-1})^i_j = \frac{C^j_i}{\det(A)}, \tag{B.7}$$

$$A^{-1} = \frac{C^T}{\det(A)}. \tag{B.8}$$

B.2 Useful formulas with matrix operations

In the following, as in the last section, we always assume $n \times n$ matrices.

The *product rule* of two matrices is as follows: Let $C = AB$. The element

c^i_j of the matrix C is formed by the i-th row of A and the j-th column of B, in which these are multiplied term by term and then added up, so

$$c^i_j = a^i_k b^k_j = \begin{pmatrix} a^i_1 & a^i_2 & \cdots & a^i_n \end{pmatrix} \begin{pmatrix} b^1_j \\ b^2_j \\ \vdots \\ b^n_j \end{pmatrix}$$

$$= a^i_1 b^1_j + a^i_2 b^2_j + \cdots + a^i_n b^n_j.$$

(B.9)

$c^i_j = a^i_k b^k_j$ means a contraction over the index k.

In this context, reference should be made to Remark 1.5 with Example 1.1. In a tensorial expression $a^i_k b^k_j$ and $b^k_j a^i_k$ are completely identical, because a^i_k and b^k_j are numbers and the multiplication of numbers is commutative. If, however, it is a question of assigning a numerical value to c^i_j, then attention must be paid to which index is used for the contraction and the arrangement of the matrices must be chosen in the calculation according to the product rule for matrices.

The *transpose* A^T of a matrix A with elements a^i_j is defined as:

$$A^T = (a^i_j)^T = a^j_i.$$

(B.10)

The *trace* of a matrix A is the sum of its diagonal entries:

$$\mathrm{tr}(A) = a^i_i = a^1_1 + \cdots + a^n_n.$$

(B.11)

Note that the trace rule is a *contraction* and cannot be carried out for a $(0,2)$-tensor, for example. Only when an index is raised by means of a metric can the trace be determined. See the comments and example in Section 4.11.

Here is a small list of important, selected matrix operations:

$$\det(A\,B) = \det(A)\det(B).$$

(B.12)

$$\det(B^{-1} A\, B) = \det(A).$$

(B.13)

$$\det\left(A^T\right) = \det\left(A\right). \tag{B.14}$$

$$\mathrm{tr}(A\,B) = \mathrm{tr}(B\,A). \tag{B.15}$$

$$\mathrm{tr}(B^{-1}\,A\,B) = \mathrm{tr}(A). \tag{B.16}$$

$$(A\,B)^T = B^T\,A^T. \tag{B.17}$$

$$(A\,B)^{-1} = B^{-1}\,A^{-1}. \tag{B.18}$$

Appendix C

Generalized Kronecker delta

C.1 Kronecker delta

The Kronecker delta is written as δ^i_j and is defined as follows

$$\delta^i_j = \begin{cases} 1 & \text{if } i = j, \\ 0 & \text{if } i \neq j. \end{cases} \tag{C.1}$$

For a vector space, e.g. of dimension $n = 3$, it can be represented as an identity matrix

$$\delta^i_j = \begin{pmatrix} 1 & 0 & 0 \\ 0 & 1 & 0 \\ 0 & 0 & 1 \end{pmatrix} = E. \tag{C.2}$$

Since the identity matrix E is the same in every coordinate system, the Kronecker delta δ^i_j is a $(1,1)$-*tensor*. This can be shown by its transformation behaviour:

$$\begin{aligned} \delta^{i'}_{j'} &= \frac{\partial x^{i'}}{\partial x^i} \frac{\partial x^j}{\partial x^{j'}} \delta^i_j \\ &= R^{i'}_i \underbrace{F^j_{j'} \delta^i_j}_{=F^i_{j'}} = R^{i'}_i F^i_{j'} = E = \delta^i_j, \end{aligned} \tag{C.3}$$

$$\implies$$

$$\delta^{i'}_{j'} = \delta^i_j.$$

A contraction over an index at two δ-terms is given by

$$\delta^i_j \, \delta^j_k = \delta^i_k. \tag{C.4}$$

For $n = 3$ one can easily check (C.4) on the basis of

$$\delta^i_j \, \delta^j_k = \delta^i_1 \, \delta^1_k + \delta^i_2 \, \delta^2_k + \delta^i_3 \, \delta^3_k,$$

by inserting the numerical values $1, 2, 3$ for the indices i and k.

There is one more thing to note that one likes to stumble upon. Based on the summation convention,

$$\delta^i_i = 3, \text{ for } n = 3. \tag{C.5}$$

C.2 Generalized Kronecker delta

The *generalized Kronecker* delta is defined as

$$\delta^{i_1 \cdots i_n}_{j_1 \cdots j_n} = \det \begin{pmatrix} \delta^{i_1}_{j_1} & \delta^{i_1}_{j_2} & \cdots & \delta^{i_1}_{j_n} \\ \delta^{i_2}_{j_1} & \delta^{i_2}_{j_2} & \cdots & \delta^{i_2}_{j_n} \\ \vdots & \vdots & \ddots & \vdots \\ \delta^{i_n}_{j_1} & \delta^{i_n}_{j_2} & \cdots & \delta^{i_n}_{j_n} \end{pmatrix} \tag{C.6}$$

The upper indices indicate the row, the lower indices indicate the columns of the matrix. The determinant of an $n \times n$ matrix has $n!$ summands, which are each a product of n matrix elements. Since δ^i_j is an $(1, 1)$-tensor, the generalized Kronecker delta is a *completely antisymmetric tensor*.

We want to deal with the generalized Kronecker delta $\delta^{ijk}_{pqr}, n = 3$, which is used in different applications. The expansion of the determinant is done via the first row:

$$\begin{aligned} \delta^{ijk}_{pqr} &= \det \begin{pmatrix} \delta^i_p & \delta^i_q & \delta^i_r \\ \delta^j_p & \delta^j_q & \delta^j_r \\ \delta^k_p & \delta^k_q & \delta^k_r \end{pmatrix} \\ &= \delta^i_p \det \begin{pmatrix} \delta^j_q & \delta^j_r \\ \delta^k_q & \delta^k_r \end{pmatrix} - \delta^i_q \det \begin{pmatrix} \delta^j_p & \delta^j_r \\ \delta^k_p & \delta^k_r \end{pmatrix} + \delta^i_r \det \begin{pmatrix} \delta^j_p & \delta^j_q \\ \delta^k_p & \delta^k_q \end{pmatrix} \\ &= \delta^i_p \, \delta^{jk}_{qr} - \delta^i_q \, \delta^{jk}_{pr} + \delta^i_r \, \delta^{jk}_{pq}. \end{aligned} \tag{C.7}$$

A contraction over the indices i and p of (C.7) (one could of course choose a different pair of indices), leads to a result that is often used in the context of the Levi-Civita symbol (see D.1 'vector triple product').

$$
\begin{aligned}
\delta^{ijk}_{iqr} &= \delta^i_i\,\delta^{jk}_{qr} - \delta^i_q\,\delta^{jk}_{ir} + \delta^i_r\,\delta^{jk}_{iq} \\
&= 3\,\delta^{jk}_{qr} - \delta^{jk}_{qr} + \delta^{jk}_{rq} \\
&= 3\,\delta^{jk}_{qr} - \delta^{jk}_{qr} - \delta^{jk}_{qr} \\
&= \delta^{jk}_{qr} \\
&= \delta^j_q\delta^k_r - \delta^j_r\delta^k_q.
\end{aligned}
\tag{C.8}
$$

And a further contraction over the indices q and j:

$$
\delta^{ijk}_{ijr} = \delta^j_j\delta^k_r - \delta^j_r\delta^k_j = 3\,\delta^k_r - \delta^k_r = 2\,\delta^k_r.
\tag{C.9}
$$

And a last contraction over r and k

$$
\delta^{ijk}_{ijk} = 2\,\delta^k_k = 2\cdot 3 = 6 = 3!.
\tag{C.10}
$$

In the following, a general rule for the reduction (contraction over a defined number of indices) of the general Kronecker delta will be derived:

$$
\delta^{i_1\,\ldots\,i_p\,i_{p+1}\,\ldots\,i_n}_{i_1\,\ldots\,i_p\,j_{p+1}\,\ldots\,j_n} = p!\,\delta^{i_{p+1}\,\ldots\,i_n}_{j_{p+1}\,\ldots\,j_n}.
\tag{C.11}
$$

The best way to derive the formula, (C.11), is based on the definition of the general Kronecker delta using a block matrix

$$
\delta^{i_1\,\ldots\,i_p\,i_{p+1}\,\ldots\,i_n}_{i_1\,\ldots\,i_p\,j_{p+1}\,\ldots\,j_n} = \det\begin{pmatrix} A & B \\ C & D \end{pmatrix},
\tag{C.12}
$$

with

$$A\,(p \times p) = \begin{pmatrix} \delta^{i_1}_{i_1} & \delta^{i_1}_{i_2} & \cdots & \delta^{i_1}_{i_p} \\ \delta^{i_2}_{i_1} & \delta^{i_2}_{i_2} & \cdots & \delta^{i_2}_{i_p} \\ \vdots & \vdots & \ddots & \vdots \\ \delta^{i_p}_{i_1} & \delta^{i_p}_{i_2} & \cdots & \delta^{i_p}_{i_p} \end{pmatrix},$$

$$B\,(p \times (n-p)) = \begin{pmatrix} \delta^{i_1}_{j_{p+1}} & \delta^{i_1}_{j_{p+2}} & \cdots & \delta^{i_1}_{j_n} \\ \delta^{i_2}_{j_{p+1}} & \delta^{i_2}_{j_{p+2}} & \cdots & \delta^{i_2}_{j_n} \\ \vdots & \vdots & \ddots & \vdots \\ \delta^{i_p}_{j_{p+1}} & \delta^{i_p}_{j_{p+2}} & \cdots & \delta^{i_p}_{j_n} \end{pmatrix},$$

$$C\,((n-p) \times p) = \begin{pmatrix} \delta^{i_{p+1}}_{i_1} & \delta^{i_{p+1}}_{i_2} & \cdots & \delta^{i_{p+1}}_{i_p} \\ \delta^{i_{p+2}}_{i_1} & \delta^{i_{p+2}}_{i_2} & \cdots & \delta^{i_{p+2}}_{i_p} \\ \vdots & \vdots & \ddots & \vdots \\ \delta^{i_n}_{i_1} & \delta^{i_n}_{i_2} & \cdots & \delta^{i_n}_{i_p} \end{pmatrix},$$

$$D\,((n-p) \times (n-p)) = \begin{pmatrix} \delta^{i_{p+1}}_{j_{p+1}} & \delta^{i_{p+1}}_{j_{p+2}} & \cdots & \delta^{i_{p+1}}_{j_n} \\ \delta^{i_{p+2}}_{j_{p+1}} & \delta^{i_{p+2}}_{j_{p+2}} & \cdots & \delta^{i_{p+2}}_{j_n} \\ \vdots & \vdots & \ddots & \vdots \\ \delta^{i_n}_{j_{p+1}} & \delta^{i_n}_{j_{p+2}} & \cdots & \delta^{i_n}_{j_n} \end{pmatrix}.$$

The numbers $\{i_1, \ldots i_p\}$ are extracted from the set $\{1, \ldots, n\}$ and therefore the following applies

$$\{i_1, \ldots i_p\} \cap \{i_{p+1}, \ldots i_n\} = \emptyset,$$
$$\{i_1, \ldots i_p\} \cap \{j_{p+1}, \ldots j_n\} = \emptyset.$$

Therefore the matrices B and C have only zero entries. And thus we obtain the following expression:

$$\delta^{i_1 \ldots i_p\, i_{p+1} \ldots i_n}_{i_1 \ldots i_p\, j_{p+1} \ldots j_n} = \det \begin{pmatrix} A & B \\ C & D \end{pmatrix} = \det(A)\det(D). \qquad (C.13)$$

For $\det(A)$ we receive

$$\det(A) = \det \begin{pmatrix} \delta^{i_1}_{i_1} & \delta^{i_1}_{i_2} & \cdots & \delta^{i_1}_{i_p} \\ \delta^{i_2}_{i_1} & \delta^{i_2}_{i_2} & \cdots & \delta^{i_2}_{i_p} \\ \vdots & \vdots & \ddots & \vdots \\ \delta^{i_p}_{i_1} & \delta^{i_p}_{i_2} & \cdots & \delta^{i_p}_{i_p} \end{pmatrix} = \det \begin{pmatrix} \delta^{i_1}_{i_1} & 0 & \cdots & 0 \\ 0 & \delta^{i_2}_{i_2} & \cdots & 0 \\ \vdots & \vdots & \ddots & \vdots \\ 0 & 0 & \cdots & \delta^{i_p}_{i_p} \end{pmatrix}$$

$$= \delta^{i_1}_{i_1} \det \begin{pmatrix} \delta^{i_2}_{i_2} & \cdots & 0 \\ \vdots & \ddots & \vdots \\ 0 & \cdots & \delta^{i_p}_{i_p} \end{pmatrix} = p \det \begin{pmatrix} \delta^{i_2}_{i_2} & \cdots & 0 \\ \vdots & \ddots & \vdots \\ 0 & \cdots & \delta^{i_p}_{i_p} \end{pmatrix}$$

$$= p\, \delta^{i_2}_{i_2} \det \begin{pmatrix} \delta^{i_3}_{i_3} & \cdots & 0 \\ \vdots & \ddots & \vdots \\ 0 & \cdots & \delta^{i_p}_{i_p} \end{pmatrix} = p\,(p-1) \det \begin{pmatrix} \delta^{i_3}_{i_3} & \cdots & 0 \\ \vdots & \ddots & \vdots \\ 0 & \cdots & \delta^{i_p}_{i_p} \end{pmatrix}$$

$$\vdots$$

$$= p\,(p-1) \cdots 2 \cdot 1 = p!.$$

And as a result we get as claimed (C.11)

$$\delta^{i_1 \ldots i_p\, i_{p+1} \ldots i_n}_{i_1 \ldots i_p\, j_{p+1} \ldots j_n} = \det(A)\,)\det(D) = p!\,\det(D) = p!\,\delta^{i_{p+1} \ldots i_n}_{j_{p+1} \ldots j_n}. \tag{C.14}$$

Appendix D

Levi-Civita symbol

The Levi-Civita symbols are defined by:

$$\epsilon_{j_1 \ldots j_n} = \delta^{1 \ldots n}_{j_1 \ldots j_n},$$
$$\epsilon^{i_1 \ldots i_n} = \delta^{i_1 \ldots i_n}_{1 \ldots n}. \tag{D.1}$$

Based on the definition of the general Kronecker delta (C.6), it follows that $\epsilon_{j_1 \ldots j_n}$ and $\epsilon^{i_1 \ldots i_n}$ are antisymmetric with respect to in the exchange of *any* two indices. To bring the index sequence, $j_1 \ldots j_n$ or $i_1 \ldots i_n$, back into the natural order $1 \ldots n$, a certain number of transpositions are necessary. If this number is even, we speak of an *even permutation* sequence, $j_1 \ldots j_n$ or $i_1 \ldots i_n$, if this number is odd, we speak of an *odd permutation* sequence (see also Box 4.1). And therefore

$$\epsilon_{j_1 \ldots j_n} = \begin{cases} +1, & j_1 \ldots j_n \text{ is an even permutation of } 1 \ldots n, \\ -1, & j_1 \ldots j_n \text{ is an odd permutation of } 1 \ldots n, \\ 0 & j_1 \ldots j_n \text{ is not a permutation of } 1 \ldots n. \end{cases} \tag{D.2}$$

$$\epsilon^{i_1 \ldots i_n} = \begin{cases} +1, & i_1 \ldots i_n \text{ is an even permutation of } 1 \ldots n, \\ -1, & i_1 \ldots i_n \text{ is an odd permutation of } 1 \ldots n, \\ 0 & i_1 \ldots i_n \text{ is not a permutation of } 1 \ldots n. \end{cases} \tag{D.3}$$

The close relationship between the Levi-Civita symbols (D.1) and the general Kronecker delta (C.6) is made very obvious by the following important expression:

$$\epsilon_{j_1 \ldots j_n} \, \epsilon^{i_1 \ldots i_n} = \delta^{1 \ldots n}_{j_1 \ldots j_n} \, \delta^{i_1 \ldots i_n}_{1 \ldots n} = \delta^{i_1 \ldots i_n}_{j_1 \ldots j_n}. \tag{D.4}$$

In Section 4.6, (4.40), we showed that the Levi-Civita symbol $\epsilon_{j_1 \ldots j_n}$ is a tensor density with the weight factor Jacobian J. With the product of the two Levi-Civita symbols, these weight factors J and J^{-1} cancel each other out and the product $\epsilon_{j_1 \ldots j_n} \epsilon^{i_1 \ldots i_n}$ is a tensor, namely the general Kronecker delta $\delta^{i_1 \ldots i_n}_{j_1 \ldots j_n}$.

Another important relationship between the Levi-Civita symbols arises in the case of a contraction via identical index sequences $i_1 \ldots i_n$, $n!/2$ even permutations and $n!/2$ permutations. Using equation (C.14) we get

$$\epsilon_{i_1 \ldots i_n} \epsilon^{i_1 \ldots i_n} = n!. \tag{D.5}$$

And the further corollary relationships:

$$\epsilon_{1 \ldots n} \epsilon^{1 \ldots n} = 1, \tag{D.6}$$

$$\epsilon_{i_1 \ldots i_p j_{p+1} \ldots j_n} \epsilon^{i_1 \ldots i_p i_{p+1} \ldots i_n} = p! \, \epsilon_{j_{p+1} \ldots j_n} \epsilon^{i_{p+1} \ldots i_n}. \tag{D.7}$$

D.1 Vector cross and triple product

It should be noted that the vector cross product exists only for a three-dimensional vector space. Just for $n = 3$ the Hodge dual of a 2-form is a 1-form, see Example 8.5.

The cross product between two vectors $\vec{v}$ and $\vec{w}$ is defined as

$$(\vec{v} \times \vec{w})^i = \epsilon^i_{\ jk} v^j w^k. \tag{D.8}$$

And for the vector triple product we get

$$\begin{aligned}
(\vec{u} \times (\vec{v} \times \vec{w}))^i &= \epsilon^i_{\ jk} u^j (\vec{v} \times \vec{w})^k \\
&= \epsilon^i_{\ jk} \epsilon^k_{\ lm} u^j v^l w^m \\
&= (\delta^i_l \delta_{jm} - \delta^i_m \delta_{jl}) u^j v^l w^m \\
&= u_m v^i w^m - u_l v^l w^i \\
&= \vec{u} \cdot \vec{w} \, v^i - \vec{u} \cdot \vec{v} \, w^i, \\
&\implies
\end{aligned} \tag{D.9}$$

$$\vec{u} \times (\vec{v} \times \vec{w}) = (\vec{u} \cdot \vec{w}) \, \vec{v} - (\vec{u} \cdot \vec{v}) \, \vec{w}.$$

Index

Bibliography

[Car19]   Sean M Carroll. *Spacetime and geometry.* Cambridge University Press, 2019.

[Dra14]   Tevian Dray. *Differential forms and the geometry of general relativity.* CRC Press, 2014.

[GPS02]   Herbert Goldstein, Charles Poole, and John Safko. Classical mechanics, 2002.

[Lam15]   Kai S Lam. *Topics in Contemporary Mathematical Physics.* World Scientific Publishing Company, 2015.

[Mat12]   Paul C Matthews. *Vector calculus.* Springer Science & Business Media, 2012.

[Nak15]   Mikio Nakahara. *Differentialgeometrie, Topologie und Physik.* Springer-Verlag, 2015.

[NPR$^+$14] Dwight E Neuenschwander, Mircea Pitici, Sheldon M Ross, Colin Adams, Nalini Joshi, Janine McIntosh, Peter Stacey, and BioInfo Summer. *Tensor Calculus for Physics.* John Hopkins University Press, 2014.

[Sch09]   Bernard Schutz. *A first course in general relativity.* Cambridge university press, 2009.

[Sch13]   Florian Scheck. *Theoretische Physik 1: Mechanik.* Springer-Verlag, 2013.

[SS80]   Bernard F Schutz and Director Bernard F Schutz. *Geometrical methods of mathematical physics.* Cambridge university press, 1980.

[Tay14]   JR Taylor. Klassische mechanik, 2014.

[TMW00]   Kip S Thorne, Charles W Misner, and John Archibald Wheeler. *Gravitation.* Freeman, 2000.